INTERSCIENCE MONOGRAPHS AND TEXTS
IN PHYSICS AND ASTRONOMY

Edited by R. E. MARSHAK

Cosmic Electrodynamics

J. H. Piddington

C.S.I.R.O.—Division of Physics
Chippendale, N.S.W., Australia

A Wiley-Interscience Publication

John Wiley & Sons, Inc.,   New York   London   Sydney   Toronto

Library of Congress Catalog Card Number: 71-88318

SBN 471 68919 X

Printed in the United States of America

10 9 8 7 6 5 4 3 2 1

Preface

The first half of this century produced modern astrophysics, with its implications for the very roots of physics. At the same time, *space science* had its genesis in theories of the solar system, distant geomagnetic field, cosmic-ray trajectories, and in studies of the earth's ionosphere and solar magnetic fields. But it was only in the second half of the century that the existence and the significance of an all-pervasive cosmic magnetic field received recognition, and that electromagnetic effects were taken into account in astrophysical problems ranging from motions in our upper atmosphere to the morphologies of planets, stars and galaxies. This explosive development was brought about by a broadening interest in astronomy and space physics in general, and by three of the largest research efforts ever undertaken in any field. These are: (1) *space research* by rockets, satellites and space probes; (2) *thermonuclear research* requiring confinement of hot plasmas in magnetic fields; and, (3) *radio astronomy* which involves the observation of electrons accelerated by magnetic and electric fields throughout the universe.

It seems that one must now accept the existence of a primeval magnetic field, pervading galaxies, intergalactic space and exercising a major measure of control over the universal plasma. Studies of the effects of this field, without which our earth itself may not have formed, make up a fascinating and increasingly important part of astrophysics.

A monograph dealing with such a fast-developing subject tends to grow out of date. However, a large part of the material has now settled into a permanent form, and is required study for students of astrophysics. In addition, unsolved problems often present a bewildering variety of possibilities and it is necessary to develop methods of attack based on the study of earlier problems. Many such problems are discussed in these chapters, and it is seen that an intuitive rather than a formal mathematical approach is required in some complex configurations of plasma and field. Again, many workers specialize in one or another cosmic region—the earth's environment, stellar atmospheres, interstellar space and so on. The various problems met in these different regions are often similar, although on different scales. For example, magnetic field annihilation in neutral sheets seems to occur in solar active regions, the geomagnetic tail and, perhaps, in galactic central regions and quasars. The acceleration of particles to relativistic energies is universal, and the Rayleigh-Taylor instability seems to be widespread. It is hoped that by listing many of the problems met in most of the regions, helpful comparisons may be possible.

Following a brief introduction to the many phenomena in which cosmic electrodynamics plays a part, the historical highlights in this field are reviewed. The principal method of investigation is described in Chapter 2; this is magnetohydrodynamics, or the study of the interaction of plasmas and magnetic fields. The following two chapters are devoted to a compilation of information of general interest. In the first, the properties of the cosmic plasmas are derived. These include the electrical conductivity and other relationships between fields and macroscopic gas parameters, and the various possible types of wave motion. The second describes electrodynamic effects of universal occurrence. These include the numerous instabilities; the generation of magnetic flux and energy; the various radio emission mechanisms; and other effects which occur in many different environments.

The remaining chapters are devoted to particular environments and their notable electrodynamic phenomena. Chapters 5 through 10 deal with the solar system. Of these, Chapters 6 through 9 are concerned partly or wholly with the earth's environment and so with geophysics. Here the observational results from numerous rockets, satellites and space probes, and from many land-based geophysical observatories, are described and the competing theories of the geomagnetic cavity and tail, or auroras and magnetic storms and Van Allen particles, are reviewed. The sun, which is mainly responsible for this great variety of terrestrial phenomena, is discussed in Chapter 5, and the solar wind and interplanetary magnetic field and cosmic-ray population in Chapter 6. Some other members of the solar system are discussed in Chapter 10, as well as theories of its origin. The objects given most attention are those which have been visited by space probes and those from which non-thermal radio emission is received. Of particular interest is Jupiter and it is hoped that, for reasons made clear in this Chapter, its magnetosphere may soon be investigated by space probes.

Chapter 11 deals with assorted galactic objects which are the sites of electrodynamic phenomena. The interstellar medium is permeated by a magnetic field which plays a vital role in control of cosmic rays and in star formation. The field is compressed in stars, where it leads to other phenomena, notably the acceleration of cosmic rays as in the Crab Nebula and a variety of radio emissions including those of pulsars.

Finally, Chapters 12 and 13 cover problems of galactic morphology and activity. A good deal of this discussion is speculative, but perhaps this is justified by the fact that most of the published work is speculative. The problems themselves are outstanding, and we appear to be on the verge of a breakthrough. It is shown that magnetic fields are important in some of the normal spiral systems as well as in radio galaxies and quasars where their effects are more obvious. In fact, it is likely that a unifying electrodynamic theory may simultaneously explain many of these mysteries.

This monograph evolved from a series of lectures given at the University of Iowa. My special thanks go to Professor James Van Allen for his hospitality, suggestions and initiation of the project, and to Jerry F. Drake for his invaluable assistance with the original lecture notes.

For reading parts or all of the manuscript, and for many helpful comments, I wish to thank Drs. S.-I. Akasofu, J. G. Duthie, R. G. Giovanelli and P. Goldreich. I also thank friends and colleagues, in addition to those specifically acknowledged in the text, for many helpful private discussions.

For supplying illustrative material I thank Dr. R. G. Giovanelli, director of CSIRO Culgoora Optical Observatory (Figure 5.3) and Dr. K. B. Mather, director of the Geophysical Institute, University of Alaska (Figure 8.5).

For permission to reproduce copyright material, I thank the editors and publishers of the following journals and books: *Astrophysical Journal*, copyright by the University of Chicago Press (Figures 4.3 and 5.2); *Journal of Atmospheric and Terrestrial Physics* and *Planetary and Space Science*, copyright by Pergamon Press Limited (Figures 9.1, 8.3b, 3.2, 7.4, 8.6, 8.9, 9.3 and 12.5); *Journal of Geophysical Research*, copyright by American Geophysical Union, (Figures 6.1, 6.2, 6.4, 9.2); *Monthly Notices of the Royal Astronomical Society*, (Figures 12.3 and 12.6); *Nature*, copyright Macmillan (Journals) Limited, (Figures 10.4 and 10.5); *Physical Review Letters*, copyright The American Physical Society, (Figure 7.8); *Earth's Particles and Fields*, ed. B. M. McCormac, Van Nostrand Reinhold Company, (Figure 7.7, Table 8.1); *The Hubble Atlas of Galaxies*, ed. A. Sandage, Carnegie Institute of Washington, (Figure 12.1).

Finally, I thank the CSIRO Executive and Dr. R. G. Giovanelli, Chief of the Division of Physics, for permission to publish this work.

J. H. PIDDINGTON
Commonwealth Scientific and Industrial Research Organization

June 1969

Contents

Cosmic Electrodynamics

CHAPTER 1

Introduction

Magnetic fields are almost ubiquitous and it is rapidly becoming clearer that they play a dominant role in the evolution of the universe. It is likely that without these fields the planets would not have formed, some stars would not have formed and even galaxies or protogalaxies may never have developed from the more tenuous primeval gas. Without magnetic fields there would be no significant story to be written and no one to write it.

Until relatively recently, say within the last two or three decades, the effects of cosmic magnetic fields and even the existence of such fields have been largely ignored. However, three of the largest research efforts ever undertaken have combined to change this situation. These are *thermonuclear research* requiring the confinement of plasmas in magnetic fields, *space research* of particles and fields near the earth, and *radio astronomy* or observations of electrons being accelerated in various ways throughout the universe. The significance of magnetic fields in the solar system, in interstellar space and beyond the galaxies is being recognized and it is now difficult to avoid accepting a primeval magnetic field which fills intergalactic space and may help shape the galaxies.

1.1 Scope

Many problems in space electrodynamics, as well as some answers, have been provided by rockets, satellites and space probes. The regions investigated range from the earth's ionosphere, magnetosphere and magnetotail to the magnetic sectors of interplanetary space and to the moon, Venus and Mars. Other problems in cosmic electrodynamics have been provided by radio, optical and x-ray observations of more distant objects. Most notable are the planet Jupiter, the sun and many other stars, the notorious Crab Nebula, our own galaxy and many external galaxies, radio galaxies and quasars. In all of these objects electrodynamic effects are observed and in some cases they are the principal factor in determining the morphology.

The understanding of these many complex patterns of magnetic field, plasma and cosmic rays requires the theoretical technique of magnetohydrodynamics or hydromagnetics (Cowling, 1957), which is the study of the motion of an electrically conducting fluid in the presence of a magnetic field. Electric currents induced in the fluid as a result of its motion modify the field; at the same time their flow in the field produces mechanical forces which modify the motion. The subject owes its peculiar interest and difficulty to this interaction between the field and the motion.

More specifically, a magnetohydrodynamic problem requires the formulation of the following equations:

(1) The equations of the electromagnetic field (Maxwell's equations).

(2) Equation of force and momentum, including the effects of electromagnetic forces as well as those met in hydrodynamics.

(3) The common hydrodynamic equations of continuity, heat transport, an equation of state and perhaps equations connecting the physical properties of the material with temperature and pressure.

Thus the problems of magnetohydrodynamics may be very difficult and some simplifying assumptions may be required. One such approach is to assume the magnetic field configuration and then determine the fluid motion and the magnetic perturbation. This procedure is valid if the magnetic perturbation is small enough but it is not the hydromagnetic method and it is dangerous in some situations.

Nor should hydromagnetics be considered as synonymous with plasma dynamics, even though there is considerable overlap. In its idealized form, the former is concerned with fluids for which the transport coefficients are simple concepts and are easily measured. In the case of a plasma, transport coefficients must be calculated by the methods of kinetic theory, usually starting with the Boltzmann equation (Section 2.4). Thus, in some situations, hydromagnetics may avoid the complexities of plasma physics. On the other hand, plasma physics can include topics in which the motions of individual particles are important, thereby removing it from the realm of fluid dynamics entirely.

1.2 Historical

The subject of cosmic electrodynamics appears to have its origins in the speculations of scientists concerning the magnetism of celestial bodies and the effects of this magnetism on charged particles approaching these bodies. Birkeland in 1896 showed experimentally that a beam of electrons projected towards a uniformly magnetized sphere would be deflected towards the polar regions of the sphere over two zones, one around each pole. He also observed a luminous ring encircling the sphere. Although the processes involved here are probably very different from the formation of the terrestrial aurora and radiation zone, the experiments provided some insight into the causes of these effects (Chapters 8 and 9). Störmer studied the phenomena mathematically by considering the motion of an isolated particle and so laid the foundations of the theory of cosmic-ray trajectories. Later applications of the single-particle method failed, however, and it is now known that with rare exceptions problems in cosmic electrodynamics must be dealt with in terms of plasmas, account being taken of the simultaneous motions of numerous particles.

Meanwhile, in 1899 Bigelow suggested that the sun might be a giant magnet. He was struck by the remarkable resemblance between solar coronal plumes, seen at the time of total eclipses, and the lines of force near a magnetized sphere. In Figure 1.1 is shown a composite sketch of such plumes extending above the north and south poles of the sun;

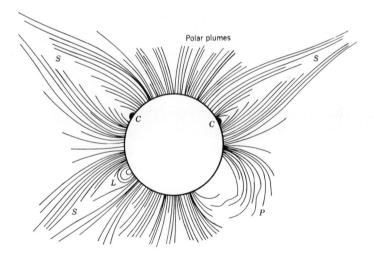

Fig. 1.1. A composite sketch of the solar corona at times of total eclipses showing a number of features which have been observed. Coronal plumes emerging from the polar regions are highly suggestive of a uniform magnetic field within the sun. Coronal streamers S and loops L are suggestive of magnetic field lines of different configurations. Other features, referred to in Section 5.1, are prominences P and condensations C.

these are indeed suggestive of magnetic control. Other features of Figure 1.1 are discussed in Chapter 5. Schuster went further and conjectured that all massive celestial bodies in rotation were great magnets and it may well be that Hale's search for solar magnetic fields was stimulated by these early speculations. In any case Hale's search was rewarded in 1908 by the discovery that sunspots were permeated by fields of the order of 1000 gauss. This discovery, and the prior knowledge that the earth was magnetized, opened up a new chapter in astrophysics and was responsible for the beginning of magnetohydrodynamics. Although the physical principles required in magnetohydrodynamics were then well established, its effects can be reproduced in the laboratory only with difficulty and are best studied in celestial bodies.

About a decade after Hale's discovery of sunspot fields, a theory of their maintenance was put forward by Larmor (1919). This was the first of the "dynamo" theories which are discussed below in Section 4.1. Larmor's theory was criticized by Cowling (1933), who added contributions to this subject and also suggested that sunspot magnetic fields might be the result of the convection of deeper-lying fields by vertical flow of plasma. This suggestion embodied the important idea of "frozen-in" magnetic fields in a conducting fluid in motion. Others were also groping towards this fundamental concept of magnetohydrodynamics. Kiepenheuer suggested that masses of solar gas ejected into the corona might carry with them parts of the surface fields. A little later Ferraro (1937) showed that in a steady non-uniformly rotating star, the angular velocity must be constant over the surfaces traced out by the magnetic field lines as they rotate. Violation would lead to the development of a toroidal field in accordance with the frozen-in principle (law of isorotation). In 1942 Alfvén clearly stated for the first time the theorem that in a highly conducting medium the magnetic lines of force are frozen into the fluid. He stated that "every motion of the liquid in relation to the lines of force is forbidden because it can give infinite eddy currents."

Meanwhile, a second line of work had been initiated which was to provide much stimulus to the theory of magnetohydrodynamics and cosmic electrodynamics. This was the study of ionized gas streams impinging on a magnetic dipole field, undertaken in connection with geomagnetic storm theory by Chapman and Ferraro (1932–1933). This early application of the hydromagnetic approach had to wait for a quarter of a century for recognition, much to the disappointment of its authors. However, it has now been developed into the generally accepted theory of geomagnetic disturbances discussed in Chapters 6–9.

Yet a third line of investigation which contributed to the early development of hydromagnetic theory and cosmic electrodynamics were the laboratory experiments with mercury and other conducting fluids (Hartmann and Lazarus, 1937, and others). The great difficulty here, compared to the astrophysical situation, was in the scale, which was so small that major astrophysical phenomena could not be reproduced.

The subjects of magnetohydrodynamics and cosmic electrodynamics began to achieve unity with the work of Alfvén (1950) starting about 1940. In addition to formulating the frozen-in concept, he discovered the simplest example of the coupling between magnetic field lines and a fluid, and showed that this interaction would produce a new kind of wave. The magnetic field lines are in tension and the plasma mass is attached to them and so a wave may travel along the lines as it does along a stretched string. Such magnetohydrodynamic waves may travel in an incompressible, as well as a compressible, fluid and so energy may be transmitted without large-scale exchanges of fluid elements as must be the case for non-conducting incompressible fluids. Alfvén also chose Faraday's method of regarding magnetic action as represented by magnetic lines of force and showed that this line of approach was most useful in a number of astrophysical problems. He developed a theory of sunspots based on hydromagnetic waves generated in the solar interior and travelling out to the surface. He introduced the idea of "guiding center" in discussing the drifts of particles gyrating in a magnetic field (Section 8.4) and this work allowed Singer to predict the existence of the Van Allen belts before their discovery. He also developed theories of geomagnetic storms and of the origin of the solar system, both based largely on electrodynamic effects. Most of this work has been collected into two books (Alfvén, 1950, 1954) which have had very considerable influence on the development of cosmic electrodynamics.

During the past decade the number of papers devoted to cosmic electrodynamics is prodigious and relatively few may be referred to in the text. Early papers described magnetic configurations capable of supporting solar gas and accounting for prominences (Chapter 5). Others discussed the pinching of electric currents by magnetic fields and the mutual annihilation of oppositely directed fields which come in contact (Section 4.2). Others described force-free magnetic fields in which the currents flow along the field lines (Section 2.2). Many were directed to an understanding of waves in a magnetized plasma, often called a magneto-ionic medium (Section 3.4). Many problems involve instabilities in magnetic plasmas, and these are discussed in the following chapter.

Meanwhile, great impetus to the study of cosmic magnetic fields developed from the rapid advances of radio astronomy after 1945.

Much of this radiation was thermal emission or bremsstrahlung from hot gas, but it was soon shown that part of the galactic radiation must be due to non-thermal processes (Piddington, 1951) which made it likely that magnetic fields were involved in one way or another (Section 4.5). Indeed Alfvén and Herlofson (1950) had already suggested that the emission from "radio stars might be by the synchrotron process." This process, which involves the acceleration of cosmic-ray electrons in magnetic fields, was also proposed by Kiepenheuer (1950). These proposals were ignored by all but the Russians Shklovskii (1953) and Ginzburg (1959) who worked out the details and made proposals which led to the establishment of the theory. It is now known that some radio galaxies have energies in the form of magnetic energy greatly exceeding their total kinetic energy. It is becoming evident that magnetic fields may shape galaxies and explode galaxies (Chapter 12) and may even play a major role in cosmology (Chapter 13). These and many other cosmic effects involving magnetic stresses have been the subject of innumerable international symposia published under the auspices of the International Astronomical Union and other bodies (see for example Pecker, 1966, STP Notes and Sturrock, 1967).

Meanwhile, the results of studies of confined plasmas undertaken in the search for controlled nuclear fusion were made public in 1957 and a tidal wave of publications was released and has grown ever since (Ware, 1961; Rose and Clark, 1961; Artsimovich, 1961; Green, 1963; International Atomic Energy Agency, 1960–1969).

Finally, the stimulus provided by the International Geophysical Year and subsequent international efforts, the wealth of data obtained from space probes, satellites and rockets, and the intensive study of astronautical problems have led to an enormous increase in interest in cosmic electrodynamics.

References

Alfvén, H., 1950, *Cosmical Electrodynamics*, Clarendon Press, Oxford.

Alfvén, H., 1954, *On the Origin of the Solar System*, Oxford University Press, London.

Alfvén, H. and Herlofson, N., 1950, *Phys. Rev.* **78**, 616.

Artsimovich, L. A., 1961, *Controlled Thermonuclear Reactions*, English Edition 1964, Oliver & Boyd, London.

Chapman, S. and Ferraro, V. C. A., 1932–1933, *Terr. Mag. Atmos. Elec.* **37**, 147; **37**, 421; **38**, 79.

Cowling, T. G., 1933, *Mon. Not. Roy. Astron. Soc.* **94**, 39.

Cowling, T. G., 1957, *Magnetohydrodynamics*, Interscience Publishers, New York.

Ferraro, V. C. A., 1937, *Mon. Not. Roy. Astron. Soc.* **97**, 458.

Ginzburg, V. L., 1959, *Paris Symposium on Radio Astronomy*, R. N. Bracewell, ed., Stanford University Press, Stanford, p. 589.

Green, T. S., 1963, *Thermonuclear Power,* Newnes, London.

Hartmann, J. and Lazarus, F., 1937, *K. Danske Vidensk. Selbskab. Mat-fys.* **15** (6) and (7).

International Atomic Energy Agency, 1960–1969, *Nuclear Fusion.*

Kiepenheuer, K. O., 1950, *Phys. Rev.* **79**, 738.

Larmor, J. L., 1919, *British Association Reports,* 159.

Pecker, J.-C., ed., 1966 *Astronomer's Handbook*, Academic Press, New York.

Piddington, J. H., 1951, *Mon. Not. Roy. Astron. Soc.* **111**, 45.

Rose, D. T. and Clark, M., 1961. *Plasmas and Controlled Fusion,* MIT Press and John Wiley & Sons, New York.

Shklovskii, I. S., 1953, *Dokl. Akad. Nauk.* **90**, 983.

STP Notes, Inter Union Commission on Solar-Terrestrial Physics c/o U.S. Nat. Academy Sci.

Sturrock, P. A., ed., 1967, *Plasma Astrophysics*, Academic Press, New York.

Ware, A. A., 1961, *Rep. Prog. Phys.* **24**, 24.

CHAPTER 2

Principles of Cosmic Electrodynamics

Most of the universe is permeated by electrically charged particles and by electromagnetic fields, and the interactions between these particles and fields constitute the subject of cosmic electrodynamics. An obvious way to study these interactions is by the motions of individual particles following the methods of Störmer, Alfvén, Northrop and Teller and others (Alfvén and Fälthammar, 1963; Northrop, 1963). Such motions provide the basis of the cooperative interaction of a group of particles, or a plasma, with the electromagnetic field. Their study also gives insight into the behavior of a plasma and so we very briefly review particle motions in the simplest cases.

Using electromagnetic units, the well-known equation of motion $\mathbf{v}$ of a particle of charge e and mass m is

$$m\frac{d\mathbf{v}}{dt} = e(\mathbf{E} + \mathbf{v} \times \mathbf{B}) \qquad (2.1)$$

where $\mathbf{E}$ and $\mathbf{B}$ are the electric and magnetic fields. When $\mathbf{B} = 0$ the particle moves with acceleration $e\mathbf{E}/m$ and so gains energy, but when $\mathbf{E} = 0$ it is accelerated in a direction perpendicular to $\mathbf{v}$ and so neither gains nor loses energy. If $\mathbf{B}$ is constant, $\mathbf{E} = 0$ and $\mathbf{v}$ is perpendicular to $\mathbf{B}$, the particle moves in a circle of radius mv/Be with angular frequency Be/m called the gyro or cyclotron frequency. If $\mathbf{v}$ has a component along $\mathbf{B}$, then the particle follows a helix of constant pitch.

If this circular motion is now perturbed by an electric field or magnetic field gradient small enough to maintain near circularity, then the particle may be regarded as gyrating about a "guiding center" which "drifts" relatively slowly. Thus if $\mathbf{E}$ and $\mathbf{B}$ are constant in space and time and perpendicular to one another, then the motion is shown in Figure 2.1, the drift velocity being $\mathbf{w} = \mathbf{E} \times \mathbf{B}/B^2$. This drift is determined by transforming to a system of axes moving with velocity $\mathbf{w}$, in which case the right-hand side of equation 2.1 reduces to $e(\mathbf{v} \times \mathbf{B})$ and the particle moves in a circle. In the presence of a magnetic field gradient

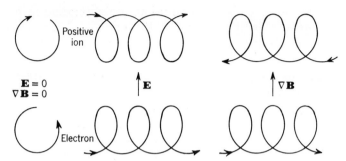

Fig. 2.1. Illustrating the motions of ions and electrons in a magnetic field **B** up from the paper. When **B** is homogeneous the particles move in circles (left) but in the presence of an electric field (center) or magnetic field gradient (right) they drift as shown.

the particle gyro radius is largest where the field is weakest and the opposite drifts of positive and negative charges may be understood by reference to Figure 2.1. These two simple drifts indicate important properties of a plasma. The first corresponds to Hall drift and Hall current while the second is the mechanism mainly responsible for the terrestrial ring current. Other drifts in more complex fields have been discussed systematically by Northrop (1963).

Clearly the study of individual particle motions gives some insight into the behaviour of an ionized gas. However, the summation of the effects of individual particles is not generally suitable as a method of studying plasmas because the combined motions constitute an electric current which changes the electric and magnetic fields. Since these fields are needed to determine particle motions we may at best resort to successive approximations. Consequently we are led to consider the combination of particles and field as a magnetoplasma which has macroscopic properties which may be used to determine simultaneous equations of field and motions. This is the magnetohydrodynamic or hydromagnetic method.

The magnetohydrodynamic equations tend to be complex but they may be simplified or idealized by making assumptions which restrict the mean free path of particles between collisions. The most stringent restriction, leading to the greatest simplification, is that the mean free path is much less than the cyclotron radius. In this case the plasma not only usually satisfies the equations of fluid dynamics but meets the additional restriction that it is isotropically electrically conducting. The corresponding equations of *idealized magnetohydrodynamics* are given in the following section and are then interpreted and discussed in Sections 2.2 and 2.3. These equations allow considerable insight into hydromagnetic processes and, in addition, are applicable to problems in a few astrophysical situations such as the interiors of stars and planets and the lower atmospheres of stars where the magnetic field strength is not excessive.

In general, the simplified approach is not adequate and we must use a more general magnetohydrodynamics of plasmas as in Section 2.4. Here the macroscopic quantities velocity, electric current density and so on are derived from the Boltzmann equation and Maxwell's field equations. In general the equations are intractable and some simplifying assumptions are necessary, although these need not be as restrictive as for the idealized equations. Two steps are available in relaxing these restrictions.

(1) The mean free path for collisions is short compared with the distances over which pressure, velocity and other macroscopic quantities change significantly. Again we have the familiar situation of fluid dynamics, where the velocity distribution is nearly isotropic and Maxwellian, but now the transport parameters and in particular the electrical conductivity are anisotropic. This means that the relationship between the electric field and current density is no longer given by the simple Ohm's law, but by a much more complicated equation.

(2) Some problems may be solved for even less restrictive conditions, when the particle mean free path is large or even infinite. A requirement is that the magnetic field is sufficiently strong so that the cyclotron radius, given above, is short compared to the distance over which all the macroscopic quantities change appreciably. In this way a group of particles moving at random is converted into a "medium" in which motions are related. However, since the magnetic field has no control of particle motions along the field lines, there is a second restriction which is that such problems must be two-dimensional, with no gradients along the field lines.

2.1 Idealized Magnetohydrodynamics

The idealized hydromagnetic equations have been listed and discussed by Cowling (1957), Alfvén and Fälthammar (1963), Ferraro and Plumpton (1966), and others. We will use electromagnetic units throughout, with magnetic permeability taken as unity. In the phenomena with which we will be concerned Maxwell's displacement current is generally small enough to be ignored. This means that all electric currents are regarded as flowing in closed circuits, and everywhere we have $\nabla \cdot \mathbf{j} = 0$, where $\mathbf{j}$ is the electric current density. This does not mean that the electrostatic effects of accumulations of charge are unimportant; on the contrary they are often very important. It only means that the current which flows to provide these accumulations is small enough to neglect. A simple and instructive exercise is to use the equations given below to prove this in the case of an astrophysical model.

The gradients of the magnetic field B are given by

$$\nabla.\mathbf{B} = 0 \qquad \nabla \times \mathbf{B} = 4\pi\mathbf{j} \qquad (2.2)$$

The total electric intensity comprises a contribution from electric space charge of density q and an induction component. These are given by

$$\nabla.\mathbf{E} = 4\pi c^2 q \qquad \nabla \times \mathbf{E} = -\frac{\partial \mathbf{B}}{\partial t} \qquad (2.3)$$

where c is the velocity of light. Finally, if we make the (temporary) simplifying assumption that the medium is isotropically conducting, then Ohm's law gives

$$\mathbf{j} = \sigma(\mathbf{E} + \mathbf{V} \times \mathbf{B}) \qquad (2.4)$$

where σ is the conductivity and $\mathbf{V}$ is the mass velocity. Combining these equations we find the "hydromagnetic" equation

$$\frac{\partial \mathbf{B}}{\partial t} = \nabla \times (\mathbf{V} \times \mathbf{B}) + \eta\nabla^2\mathbf{B} \qquad (2.5)$$

where $\eta = (4\pi\sigma)^{-1}$.

At this stage we must introduce the hydrodynamic equations, modified to take account of the interaction between the motion and the magnetic field. The equation of motion or momentum contains the usual hydrodynamic terms together with an electromagnetic or Lorentz force of density $\mathbf{j} \times \mathbf{B}$, so that

$$\frac{d\mathbf{V}}{dt} = \mathbf{F} - \rho^{-1}\nabla p + \rho^{-1}(\mathbf{j} \times \mathbf{B}) + \nu\nabla^2\mathbf{V} \qquad (2.6)$$

where $\mathbf{F}$ is the body force per unit mass (perhaps gravitational), ρ is the mass density, p is the pressure, ν is the kinematic viscosity and d/dt is the mobile operator $(\partial/\partial t) + \mathbf{V}.\nabla$. The remaining hydrodynamic equations depend on simplifying assumptions made. For example, we might assume an incompressible fluid in which case the equations present the least difficulty. Even with this drastic simplification, the hydromagnetic solutions give some physical insight into some problems in astrophysics. In general, however, we must consider the medium as compressible and add the gas equations.

The additional hydrodynamic equations needed are the equation of continuity,

$$\frac{\partial \rho}{\partial t} + \nabla.(\rho \mathbf{V}) = 0 \qquad (2.7)$$

and a relationship between pressure and density. If changes are adiabatic, then this relationship has the simple form

$$\frac{d}{dt}(p\rho^{-\gamma}) = 0 \qquad (2.8)$$

with $\gamma = 5/3$ in a fully ionized gas.

In other situations account must be taken of heat transfer by using the heat equation,

$$\frac{dW}{dt} = \frac{p}{\rho}\frac{dp}{dt} + \epsilon \qquad (2.9)$$

where W denotes the heat energy per unit volume and ϵ is the total heating effect per unit volume due to heat conduction, Joule heating and viscous dissipation of macroscopic energy. In ionized gases the most important contribution to ϵ is usually that of thermal conductivity which makes a contribution to ϵ given by

$$\epsilon' = \lambda \nabla^2 T \qquad (2.10)$$

where T is the temperature and λ is the thermal conductivity. Sometimes, however, in partially ionized gases the Joule heating is often most important and must be calculated from the equations given in Section 3.3.

2.2 Interpretation

The interpretation of these equations is generally considered to be well known, although there are subtleties which we will discuss later. The electric field used in Ohm's law has two components, one of which derives from the motion $\mathbf{V}$ of the plasma. In the following section we discuss moving magnetic field lines and field configurations, and if such motion is allowed then we must specify that the motion $\mathbf{V}$ is relative to the lines or configuration. For example, a particle experiences no acceleration because of its motion through space with the earth and the earth's magnetic field. When $\mathbf{V} = 0$ and the magnetic configuration is not changing, then the curl of $\mathbf{E}$ is also zero and the residual field has a divergence only; it must be due to space-charge accumulation.

The effects which make up a hydromagnetic phenomenon are shown schematically in Figure 2.2. Suppose that we start with a fluid motion $\mathbf{V}$

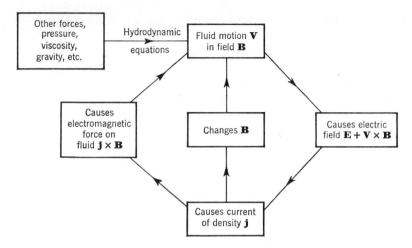

Fig. 2.2. The sequence of events in a magnetohydrodynamic phenomenon, starting with fluid motion **V** in a magnetic field **B**. The electric field induced is **V** × **B**, and in the presence of boundaries or inhomogeneities a space-charge field **E** also develops. The current **j** affects the fluid motion and also changes the magnetic field (through the curl of **B** equation) and this in turn changes the electric field. The current also changes the fluid motion (through the Lorentz force).

which causes an induction field **V** × **B** and, because of space-charge accumulations on boundaries, a total field **E** + **V** × **B** as shown on the right-hand side. The second step is the development of the overall current–density distribution **j** according to Ohm's law (equation 2.4). This current in turn changes the magnetic field according to Maxwell's equation for the curl of **B**, and during this change a new electric field component is introduced through the equation for the curl of **E** (equation 2.3). These changes are shown by the vertical arrows. The next sequential effect results from the mechanical force of density **j** × **B** which changes the fluid motion according to the force–momentum equation 2.6. Other changes may result from hydrodynamic forces as indicated by the box at top left.

The electromagnetic force density **j** × **B** is most simply interpreted in terms of Maxwell's stresses. Using equation 2.2 we have

$$\mathbf{j} \times \mathbf{B} = -\nabla (B^2/8\pi) + \frac{1}{4\pi}(\mathbf{B}.\nabla)\mathbf{B} \tag{2.11}$$

which implies that the force **j** × **B** is equivalent to a hydrostatic pressure $B^2/8\pi$, together with a tension $B^2/4\pi$ along the lines of force. This is equivalent to a tension $B^2/8\pi$ together with an equal pressure transverse to the lines, which is the form usually quoted for the Maxwell stresses. The tension endows the medium with a sort of rigidity and, in fact, waves may travel not only perpendicular to the field, but along the field with plasma motions transverse to the field lines like waves on a stretched string.

The concept of a magnetic field being frozen into a plasma is well known, but as shown by the hydromagnetic equation 2.5 freezing-in is not complete. We can investigate two extreme cases distinguished by which term on the right dominates. The two motions are separated if we assume a stationary medium; we then have the equation of diffusion and its approximate solution,

$$\frac{\partial B}{\partial t} = \eta \nabla^2 \mathbf{B} \qquad\qquad t \sim 4\pi\sigma L^2 \qquad\qquad (2.12)$$

where L is a characteristic dimension and t is the time for the field to diffuse out of the plasma. In the case of a copper sphere of 1 meter diameter the time is about 10 sec. In most astrophysical situations, however, the time is long compared with other characteristic times and the field may be considered frozen into the plasma. In the solar corona, for example, the conductivity is approximately 10^{-5} emu and the scale is 10^5 km so that the diffusion time is about 10^8 years. When freezing-in is complete, equation 2.5 reduces to

$$\frac{\partial \mathbf{B}}{\partial t} = \nabla \times (\mathbf{V} \times \mathbf{B}) \qquad\qquad (2.13)$$

which means that the magnetic flux through a closed circuit moving with the fluid is constant.

Equation 2.13 must not be always accepted in this simple sense. Plasma and field may be separated in a number of ways, a fact which is only too well known to those engaged in experiments directed at controlled nuclear fusion. Most commonly, separation is brought about by space-charge fields which may develop in a variety of ways and appear to constitute the greatest single difficulty in confinement problems. Separation is also made possible by the fact that plasma may move freely along magnetic field lines, thus leading to Rayleigh-Taylor instability and other effects discussed in later sections. In general, however, there is a strong overall tendency for cosmic magnetic fields to be frozen into cosmic plasmas and, conversely, if they are absent in the first place, then they are frozen out of the plasma and so may only develop by growth within the plasma.

Before discussing a number of problems in cosmic magnetohydrodynamics it is instructive to consider some simpler problems in which the plasma is either at rest or moving uniformly. The equation of motion 2.6 then reduces to

$$\nabla p - \rho \mathbf{g} = \frac{1}{4\pi} (\nabla \times \mathbf{B}) \times \mathbf{B} \qquad (2.14)$$

where $\mathbf{g} = \mathbf{F}$ is the gravitational force and viscosity is not considered. In these examples the Maxwell's stresses are balanced by hydrostatic forces and the cases where the plasma is stationary are *magnetohydrostatic* problems. One important class of these problems is that of the stability of magnetostatic configurations, a requirement for plasma containment. Another important class is for *force-free magnetic fields*, in which case the Maxwell's stresses balance one another and the hydrostatic forces are zero.

A simple example of magnetohydrostatic balance is found in sunspots which are about 10^3K cooler than the surrounding photosphere. Being cooler, they should be denser and so should sink and disappear in a matter of minutes whereas they persist for weeks. The answer, as suggested by Alfvén, is in terms of magnetic pressure $B^2/8\pi$ exerted laterally as shown in Figure 2.3a. The gas pressure p in the magnetic cylinder must be lower by this amount than the pressure outside the cylinder at the same level. Since B is independent of height z (along

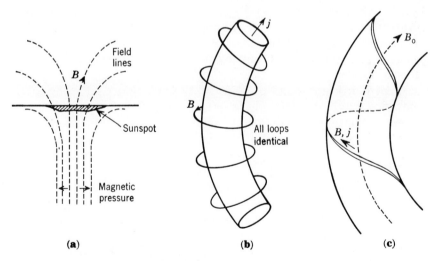

Fig. 2.3. Some magnetohydrostatic phenomena. (a) A cylindrical sunspot magnetic field (dashed lines) tends to expand outwards and so lowers the pressure and temperature below the spot. (b) A current pinch is unstable because a slight bend weakens the field on the convex side of the plasma and so leads to a sharper bend. (c) A schematic representation of a force-free magnetic field, described in the text.

the cylindrical part of the field), $\partial p/\partial z$ is the same inside and outside the cylinder. Also, since $\partial p/\partial z = -\rho g$, the same is true for the density and since $T \propto p/\rho$, the temperature, like the pressure, must be lower inside the cylinder than outside. It is found that fields of 1000 gauss or more should have substantial cooling effects and that near the photosphere they are too strong to be contained as a cylinder and burst out as shown. At a depth of a few thousand kilometers, however, the gas pressure exceeds the magnetic pressure by factors of about 1000 and the cooling effect is negligible. Thus the spot is essentially a surface cooling phenomenon. Other problems of solar magnetohydrostatics are mentioned in Chapter 5 and those of galactic magnetohydrostatics are discussed in Chapters 11 and 12.

The preponderance of work on magnetohydrostatic stability has been done in connection with thermonuclear devices, whose development has been a continuous struggle against instabilities (Ware, 1961). Some of these instabilities occur in cosmic situations:

(1) The kink instability arises when a plasma carrying a line current j as shown in Figure 2.3b becomes slightly bent. The lines of force are brought closer together on the concave side and separated on the convex side. Thus the increased field on the concave side exerts more pressure to accentuate the pinch.

(2) Rayleigh-Taylor instability is well known in hydrodynamics, occurring when a layer of heavy liquid rests on a layer of lighter liquid. The former "drips" down through the latter. The hydromagnetic analogue is introduced in Chapter 4 and discussed in several situations in later chapters.

(3) Exchange instability depends on the fact that parallel magnetic field lines may move freely past one another carrying their attached plasma particles and without changing their shapes. Plasma with a weak field changes place with plasma with a strong field and may reduce the total energy of the system; if so, then energy is available to drive the exchange motions. These are discussed in connection with magnetospheric convection near the earth and Jupiter.

(4) Other instabilities are the sausage instability and the mirror instability. Suppose that a local decrease of radius occurs in the current pinch of Figure 2.3b. The lines of force in the constriction are crowded together and the field becomes stronger and so is able to reduce the radius even further. This is the "sausage" instability, and in extreme cases the magnetic loops may shrink to points and stop current flow. Anisotropy instabilities occur when the plasma pressure $p_\perp$ across the field lines is too large ("mirror" instability) or too small ("firehose"

instability) compared with the pressure $p_{\parallel}$ along the lines (see Spitzer, 1962, p. 106).

In general, there is a tendency to hydromagnetic stability when the curvature of the magnetic lines of force is towards regions of lower plasma density, or when the plasma is concave to the magnetic field. Stability is also increased by making the field configuration as force-free as possible: magnetic stresses tend to balance and so to remove some forms of instability.

A special case of the magnetohydrostatic equation 2.14 occurs when the net hydrostatic forces are zero and so the magnetic field is force-free. This must occur of course in a non-conductor where $\nabla \times \mathbf{B} = 0$, but it may also occur where currents flow, provided they flow along the field lines so that

$$(\nabla \times \mathbf{B}) \times \mathbf{B} = 0 \qquad i.e., \ \nabla \times \mathbf{B} = \alpha \mathbf{B} \qquad (2.15)$$

where α is a scalar function of position. Such fields are likely to be important in nature because when the field is strong enough to overwhelm the plasma the Maxwell stresses must balance one another according to equation 2.15. However, no field configuration may be entirely force-free as this requires infinite extent and infinite magnetic energy. There must be some boundary where mechanical forces restrain the field and in the case of sunspot fields these mechanical forces are applied below the photosphere where gas pressure greatly exceeds magnetic pressure.

Solutions to the force-free equation 2.15 have been discussed by Woltjer (1957), Chandrasekhar and Kendall (1957) and others. They may take quite complicated forms, but we will illustrate their general nature by a simple cylindrical example. In a system r, θ, z of cylindrical polar coordinates assume a field with only two components $B_\theta(r)$ and $B_z(r)$. It is obvious at once that such a field could not be force-free at its ends, but this difficulty is overcome if the long cylinder is bent into a circular form with ends joined. The vector equation yields one component equation,

$$\frac{d}{dr}(B_\theta{}^2 + B_z{}^2) + \frac{2}{r}B_\theta{}^2 = 0 \qquad (2.16)$$

showing that the magnetic intensity $(B_\theta{}^2 + B_z{}^2)^{1/2}$ decreases steadily as r increases. Suppose further that $\alpha r B_z = B_\theta$, where α is now a constant which represents a uniform twist of the lines of force. Then we find from equation 2.16 that

$$B_\theta = B_0 \alpha r \{1 + (\alpha r)^2\}^{-1} \qquad\qquad B_z = B_0 \{1 + (\alpha r)^2\}^{-1} \quad (2.17)$$

where $B_o = B_z(r = 0)$ is the axial field at the center. The current lines coincide with the lines of force and both are shown schematically in Figure 2.3c. The total magnetic flux along the cylinder is constant and if we denote this by ψ, then the field strength in the center is given by

$$B_o = (a^2\psi/\pi) \, 1n(1 + a^2R^2) \qquad (2.18)$$

where R is the radius of the cylinder. Hence the greater the twist parameter a, the greater the concentration of axial field near the center. The tension in the lines of force wrapped around the cylinder compresses the inner lines, as would happen in a cable of twisted rubber bands.

2.3   Moving Magnetic Field Lines

Our first encounter with a magnetic field was probably that caused by current flowing in a piece of wire, the field being a secondary effect. A field frozen into a plasma, on the other hand, is determined by the connection of the lines of force with the plasma as in equation 2.13. If the plasma undergoes deformations, the electric currents are determined, not by Ohm's law which has little significance, but by equation 2.2. The magnetic field is the *primary effect*. As long as field lines are immersed in a plasma they may be regarded as entities which may be bent, stretched, and moved bodily.

The concept of moving lines of force has been of great help in the intuitive approach to space problems. It derives from the frozen-in equation 2.13 and the theory parallels earlier work on vortex lines in ideal fluids. These satisfy the basic equation of angular velocity Ω,

$$\frac{\partial \Omega}{\partial t} - \nabla \times (\mathbf{V} \times \Omega) = 0 \qquad (2.19)$$

which is formally identical with the frozen-in equation 2.13. Stern (1966) has discussed moving field lines from this point of view.

However, this is not the full story, although it suffices in many astrophysical problems. In addition to the diffusion of field lines permitted by finite electrical conductivity as given by the hydromagnetic equation 2.5, relative motion of plasma and field lines may result from two effects not revealed in this equation.

(1) As shown in Section 3.2, the magnetic lines of force tend to convect with the electrons rather than the ions.

(2) Electric space-charge fields allow relative motion of the whole plasma across the field, or motion of field lines across the plasma.

Since such fields have divergence but no curl, they appear in the first but not in the second of equations 2.3, nor in the frozen-in equation 2.13—they can have no effect on the uniform expansion or contraction of a bundle of field lines or tube of force. However, they do appear in the equation which states that the net electric field experienced by the perfectly conducting plasma is zero. In the absence of acceleration or pressure gradients, this is

$$\mathbf{E} + \mathbf{V} \times \mathbf{B} = 0 \qquad (2.20)$$

Depending on the past history of the plasma and on boundaries, $\mathbf{E}$ and $\mathbf{V}$ may have any values at all provided they satisfy this equation. For example, Chapman and Ferraro have analyzed the problem of an infinite plane slab of plasma escaping across the solar magnetic field. If the slab is polarized, then the particles (except near the boundary) move in straight lines and escape from the sun. An individual particle could not do so.

An interesting example of equation 2.20 is its application to a rotating magnetized sphere and plasma. We may regard the external plasma and the sphere as rotating relative to the stationary field lines and an electric potential field developing which satisfies equation 2.20. Alternatively, we may regard the field lines as moving with the sphere so that in this rotating system we have $\mathbf{E} = \mathbf{V} = 0$. This situation obtains provided that the plasma density is everywhere sufficient to cancel the field $\mathbf{V} \times \mathbf{B}$ and to ensure that the field along the magnetic field lines is zero. If the space outside the conducting sphere is a vacuum, then the motion of an individual particle, in general, bears no relation to field line motion. These different situations have been discussed by Deutsch (1955), Birmingham and Jones (1968) and others.

In the case of the earth's magnetic field we have a curious, inter- mediate situation. In and above the ionosphere there is adequate plasma to ensure motion of the plasma with the field lines. However, below the ionosphere the gas is non-conducting and cannot cancel out electric fields along the magnetic field lines. Thus a particular field line projecting down through the ionosphere cannot be identified with any line projecting through the surface of the earth. There is "slipping" between the two systems as described in Chapters 7 and 8.

2.4 Magnetohydrodynamics of Plasmas

In some cosmic situations the approximations of idealized magneto- hydrodynamics are not valid: We may not treat the plasma as a fluid but must take into account the velocity distributions within each of the two or more particle populations which make up the plasma. Macroscopic quantities are then determined by integrating over velocity space.

We start with the quantity f, the density of particles in phase space as a function of position $\mathbf{r}$ and velocity $\mathbf{v}$. Then $f(\mathbf{r},\mathbf{v},t)dxdydzdv_xdv_ydv_z$ is the number of particles within the spatial volume $dxdydz$ centered at $\mathbf{r}$, and whose velocities lie within the intervals dv_x, dv_y and dv_z centered at $\mathbf{v}$. The differential Df/Dt is defined as the rate of change of f along the *free trajectory* of a particle, collisions between particles being ignored in the computation of this trajectory. This differential, its expansion and its application have been discussed in detail by Chapman and Cowling (1953).

In a conservative system, changes in f result only from collisions between particles, and so in the absence of collisions we have $Df/Dt = 0$. This is Liouville's theorem which states that f is constant along a dynamical trajectory.

When collisions are significant we must introduce a term $J(f,f)$ which represents the change in f of a particular population resulting from collisions at the given point in space and time with particles of other populations. Contributions to J may result from physical encounters and by distant electromagnetic interactions. When these are taken into account we have $Df/Dt = J(f,f)$.

If the differential is now expanded we have Boltzmann's equation,

$$\frac{\partial f}{\partial t} + \mathbf{v}.\nabla f + \frac{\mathbf{F}}{m}\cdot\frac{\partial f}{\partial \mathbf{v}} = J(f,f) \qquad (2.21)$$

where a particle of mass m experiences a total force $\mathbf{F}$, including an electromagnetic force $e(\mathbf{E} + \mathbf{v} \times \mathbf{B})$ as well as gravitational force. In the case where the plasma is sufficiently hot and tenuous so that particle collisions may be neglected we have the collisionless Boltzmann equation; combined with Maxwell's field equations they become the set of Vlasov equations. Discussions of these various equations, their applications and solutions have been given by Drummond (1961), Spitzer (1962), Thompson (1962), Longmire (1963), Montgomery and Tidman (1964) and Schmidt (1966).

The solution of equation 2.21 is generally complicated, and here we will only indicate the method used to derive the basic macroscopic equations. First, the macroscopic particle density $n(\mathbf{r},t)$ is given by the integral in velocity space.

$$n(\mathbf{r},t) = \int\limits_{-\infty}^{+\infty} \int \int f(\mathbf{r},\mathbf{v},t)\, dv_x dv_y dv_z \qquad (2.22)$$

Other macroscopic quantities $Q(\mathbf{r},t)$, such as momentum density or current density, may be found from the equation

$$Q(\mathbf{r},t) = \int\limits_{-\infty}^{+\infty} \int \int q(\mathbf{v})\, f(\mathbf{r},\mathbf{v},t)\, dv_x dv_y dv_z \qquad (2.23)$$

where $q(\mathbf{v})$ is the appropriate function. Quantities which are means for all particles of a population are found by dividing the integral by n. Thus if we put $q(\mathbf{v}) = \mathbf{v}$ we find the macroscopic velocity.

$$\mathbf{V}(\mathbf{r},t) = \frac{1}{n(\mathbf{r},t)} \int\limits_{-\infty}^{+\infty} \int \int \mathbf{v} f(\mathbf{r},\mathbf{v},t)\, dv_x dv_y dv_z \qquad (2.24)$$

Relations between macroscopic quantities may be found by multiplying equation 2.21 by $q(\mathbf{v})\, dv_x dv_y dv_z$ and integrating over all $\mathbf{v}$. If $q = 1$, this leads to the equation of continuity.

$$\frac{\partial n}{\partial t} + \nabla.(n\mathbf{V}) = 0 \qquad (2.25)$$

The equation of momentum transfer may be obtained by putting q equal to $m\mathbf{v}$. The total force on a particle is taken as

$$\mathbf{F} = e(\mathbf{E} + \mathbf{v} \times \mathbf{B}) - m\nabla\phi \qquad (2.26)$$

where ϕ is the gravitational potential. The required equation is then (Spitzer (1962), p. 157)

$$nm\left(\frac{\partial \mathbf{V}}{\partial t} + \mathbf{V}.\nabla\mathbf{V}\right) = ne(\mathbf{E} + \mathbf{V} \times \mathbf{B}) - \nabla.[\psi] - nm\nabla\phi + \mathbf{P} \qquad (2.27)$$

where $[\psi]$ is the hydrodynamic stress tensor and $\mathbf{P}$ is the net rate of momentum gain due to collisions of the group of particles of mass m and charge e. A separate equation of this type is required for each type of particle, so that at least two such equations are needed. The stress tensor is defined in terms of the mean velocity $\mathbf{V}$ of the group and the random velocity $\mathbf{v}$ of individual particles, so that

$$\mathbf{V} = \frac{1}{n}\, \Sigma\mathbf{v} \qquad\qquad [\psi] = m\, \Sigma(\mathbf{v} - \mathbf{V})\,(\mathbf{v} - \mathbf{V}) \qquad (2.28)$$

where in each case summation is over unit volume. In general the stress tensor introduces such complexity that the macroscopic equations become unusable. However, in some situations it may be replaced by the scalar pressure p so that $\nabla.[\psi] = \nabla p$. These situations are when the mean free path between collisions is short compared to the distances over which $\mathbf{V}$ and other macroscopic quantities change appreciably, and for long free paths provided the radius of gyration in the magnetic field is correspondingly short and provided also that all gradients along the field lines may be neglected (the two-dimensional case).

For a binary gas comprising equal numbers of electrons and ions of mass m, M and charge $\mp e$ we have two equations of momentum and adopting the above simplification these together yield the macroscopic quantities, velocity $\mathbf{V}$ of the whole plasma and current density $\mathbf{j}$.

$$\mathbf{V} = n\rho^{-1}(M\mathbf{V}_i + m\mathbf{V}_e) \tag{2.29}$$

$$\mathbf{j} = en(\mathbf{V}_i - \mathbf{V}_e) \tag{2.30}$$

Further simplification may be made by linearizing the equations, and if we then add $nM(\partial \mathbf{V}_i/\partial t)$ and $nm(\partial \mathbf{V}_e/\partial t)$ we have the equation of motion.

$$\rho\frac{\partial \mathbf{V}}{\partial t} = \mathbf{j} \times \mathbf{B} - \nabla p - \rho\nabla\phi \tag{2.31}$$

Subtracting $n(\partial \mathbf{V}_e/\partial t)$ from $n(\partial \mathbf{V}_i/\partial t)$ we have

$$\mathbf{E} + \mathbf{V} \times \mathbf{B} - \sigma_0^{-1}\mathbf{j} + (e\rho)^{-1}$$
$$\{M\nabla p_e - m\nabla p_i - (M - m)\mathbf{j} \times \mathbf{B}\} = 0 \tag{2.32}$$

where a term in $\partial \mathbf{j}/\partial t$ has been omitted and the resistivity term is defined by $\mathbf{P} = \sigma_0\, ne\mathbf{j}$. This relationship (equation 2.32) between electric field and current is sometimes referred to as the "generalized Ohm's law" and when $\mathbf{B}$, ∇p_i and ∇p_e all vanish it reduces to Ohm's law with σ_0 equal to the electrical conductivity.

The accurate determination of σ_0 is complicated and it is found that it is about twice as large for currents flowing parallel to $\mathbf{B}$ as for currents perpendicular to a strong field (Spitzer, 1962). For the present purposes, however, a simple derivation is adequate based on the momentum exchange between ions and electrons per unit volume and time.

$$\mathbf{P} = nm\tau^{-1}(\mathbf{V}_i - \mathbf{V}_e) \tag{2.33}$$

Here τ is the mean period between collisions. From equation 2.30 and our definition of σ_0 we find

$$\sigma_0 = \frac{ne^2\tau}{m} \tag{2.34}$$

This is the electrical conductivity of a binary gas in the absence of a magnetic field and also the measure of dissipation for current flowing in any direction.

The generalized Ohm's law in the form used below neglects terms involving m/M and so equation 2.32 becomes

$$\mathbf{E} + \mathbf{V} \times \mathbf{B} = \sigma_0{}^{-1}\mathbf{j} - (ne)^{-1}(\nabla p_e - \mathbf{j} \times \mathbf{B}) \tag{2.35}$$

References

Alfvén, H. and Fälthammar, C. G., 1963, *Cosmical Electrodynamics*, Oxford University Press, London.

Birmingham, T. J., and Jones, F. C., 1968, *J. Geophys. Res.* **73**, 5505.

Chandrasekhar, S., 1961, *Hydrodynamic and Hydromagnetic Stability*, Clarendon Press, Oxford.

Chandrasekhar, S., and Kendall, P. C., 1957, *Astrophys. J.* **126**, 457.

Chapman, S. and Cowling, T. G., 1953, *The Mathematical Theory of Non-uniform Gases*, Cambridge University Press, Cambridge.

Cowling, T. G., 1957, *Magnetohydrodynamics*, Interscience Publishers, New York.

Deutsch, A. J., 1955, *Ann. d'Astrophys.* **18**, 1.

Drummond, J. E., ed., 1961, *Plasma Physics*, McGraw-Hill, New York.

Ferraro, V. C. A., and Plumpton, C., 1966, *Magneto-Fluid Dynamics*, 2nd ed., Oxford University Press, London.

Longmire, C. L., 1963, *Elementary Plasma Physics*, Interscience Publishers, New York.

Montgomery, D. C., and Tidman, D. A., 1964, *Plasma Kinetic Theory*, McGraw-Hill, New York.

Northrop, T. G., 1963, *The Adiabatic Motion of Charged Particles*, Interscience Publishers, New York.

Schmidt, G., 1966, *Physics of High Temperature Plasmas*, Academic Press, New York.

Spitzer, L., 1962, *Physics of Fully Ionized Gases*, Interscience Publishers, New York.

Stern, D. P., 1966, *Space Sci. Rev.* **6**, 147.

Thompson, W. B., 1962, *An Introduction to Plasma Physics*, Pergamon Press, London.

Ware, A. A., 1961, *Rep. Progr. Phys.* **24**, 24.

Woltjer, L., 1957, *Bull. Astron. Inst. Netherlands* **14**, 39.

CHAPTER 3

Cosmic Plasmas

All but an infinitesimal part of the universe appears to be made up of plasma, probably all of it magneto-plasma. Thus in most problems in cosmic electrodynamics we are dealing with a non-rigid and electrically conducting medium extending to infinity. This situation is quite different from those of laboratory electrodynamics, where experiments involve bodies of relatively small dimensions and bodies which are rigid or non-conducting. As far as we know the latter type of environment is confined to the surfaces of planets and the interiors of satellites such as the moon.

The remainder of the universe is made up of various plasmas, and Table 3.1 shows the enormous variations in the dimensions, density, temperature and degree of ionization of these plasmas. Their mass densities vary from about 10^{-29} g/cm^3 or less between the distant galaxies (Section 13.4) to an average value of about unity in the sun,

Table 3.1 Cosmic Plasmas

Region	Extent	Mass density, g/cm^3	Temperature, K	Proportion ionized
Ionosphere	1 R$_E$*	10^{-20}–10^{-10}	200–1500	10^{-9}–1
Magnetosphere	10–10^3 R$_E$	10^{-21}	10^4	1
Sun (stars)	10^6 km	1.4 (mean)	10^4–10^7	10^{-4}–1
Solar corona	10^7 km	10^{-19}–10^{-16}	10^6	1
Solar system	100 AU†	10^{-23}	10^5	1
Galactic nebulae	1–100 pc**	to 10^{-20}	10^2–10^4	10^{-4}–1
Galaxy	30 kpc	10^{-24}	10^2–10^4	10^{-4}–1
Local cluster	1 Mpc	10^{-27}	10^5?	1
Universe	10^4 Mpc	10^{-29}	10^5–10^6?	1

* R$_E$ or earth radius of 6380 km.
† AU or astronomical unit of 1 $\times 10^{13}$ cm.
** Parsec or 3.1×10^{18} cm.

up to about 10^8 g/cm^3 in white dwarf stars and very much higher in neutron stars. The temperatures range from about 100K or less in the dark interiors of HI regions (Section 11.1) to tens of millions in the centers of some stars. The plasmas with the lowest degrees of ionization are the lower parts of the ionospheres of the earth and other planets. These regions are really boundary regions between the universal plasma and the small insulating, atmospheric shells. In the earth's ionosphere the proportion of ionization falls to 10^{-13} or less at the very lowest levels. In interstellar HI regions and near the surfaces of some stars the proportion is about 10^{-4}. This is provided mainly by metal atoms with low ionization potentials and although small is still enough to render the gas electrically conducting.

A very important basic property of plasmas is their tendency towards electrical neutrality. Macroscopically, if the number densities of electrons and ions differ appreciably, then the electrical potentials of some particles greatly exceed their thermal energies and unless some special mechanism is available they will move rapidly to restore neutrality. For example, suppose that in a plasma we have a region in the form of a plane slab of thickness $2h$ from which all electrons have been removed to leave a space charge of density $q = ne$, where n, e are the ion density and charge. The field gradient (equation 2.3) is $4\pi nec^2$ and the work done moving the last electron from the center of the slab to its boundary is $2\pi ne^2c^2h^2$. The mean kinetic energy of a particle in one direction is $\frac{1}{2}kT$, where k is Boltzmann's constant and T is the temperature, and if this just suffices to remove the last electron, then

$$h = \left(\frac{kT}{4\pi c^2 ne^2}\right)^{1/2} = 6.9 \left(\frac{T}{n}\right)^{1/2} \tag{3.1}$$

This is called the "Debye shielding distance" and gives a rough measure of the distance over which electron and ion densities may differ appreciably in the absence of a non-thermal source of energy. It is also a measure of the distance beyond which the field of any electric charge is shielded by the effects of other particles. In the solar corona and interplanetary space its values are about 1 cm and 10 meters, respectively. If h is small compared with other lengths of interest, then we may, following Langmuir, refer to the group of particles as a plasma.

From time to time suggestions have been made that some phenomena are explicable in terms of electric charges on cosmic bodies such as the sun and moon. These suggestions are worthless unless backed by a very definite mechanism for maintaining the charge. There might appear to be exceptions in the case of planets with insulating atmospheres, such as the earth. However, above these atmospheres we again encounter plasma (the ionosphere) on whose inner boundary sufficient space charge will accummulate to neutralize that on the surface of the planet. Further from the planet the electric field is zero.

The present chapter is devoted to further discussion of the cosmic plasmas of Table 3.1. They may conveniently be divided into "fully ionized gas," in which any neutral atoms which may be present have negligible physical effect, and partially ionized gas. The electrodynamic properties of these two types of gas differ considerably.

A relationship of basic importance is that between the electric field experienced by the plasma and the current density. In simple isotropic fluids this is Ohm's Law (equation 2.4) and the conductivity is a scalar quantity. In general the relationship is much more complicated and the current is determined not only by the field, but also by pressure gradients and inertial forces. Even in these gases the concept of conductivity components is useful in understanding some problems. They are discussed in the following section, being compared later with the more accurate $\mathbf{E}$–$\mathbf{j}$ relationship.

Another very important property of a plasma is the group of perturbations which it will permit and which it will transmit as waves. This property, discussed in Section 3.4, is very different from the phenomena which give rise to waves. The question of wave generation is discussed in Chapter 4.

We conclude the chapter with a brief discussion of the terrestrial ionosphere. It might be argued that this is only one of many cosmic plasmas and hardly warrants such special recognition. The ionosphere is more than that, however. It has been the major laboratory for plasma research for more than half a century and is still the subject of intensive investigation.

3.1   The Conductivity Components

The properties of plasmas may be studied by using equation 2.24 to determine the macroscopic velocities of each of the components, ions, electrons and neutral atoms. However, it is usually advantageous to make a direct determination of the useful macroscopic quantities $\mathbf{V}$, the mass velocity, and $\mathbf{j}$, the current density. For a binary gas with electrical neutrality these are given by equations 2.29 and 2.30. The two basic equations involving these quantities are then the equation of motion (equation 2.31) and the relationship between the electric field and the current density, or generalized Ohm's law (equation 2.35).

In some plasmas the $\mathbf{E}$–$\mathbf{j}$ relationship may be written in terms of a transport parameter, the conductivity σ. When the collision frequency is much greater than the cyclotron frequency the medium is isotropically conducting and σ is a scalar as in equation 2.4. For lower collision frequencies, the conductivity becomes a tensor and we have

$$\mathbf{j} = \sigma_0 \mathbf{E}'_\parallel + \sigma_1 \mathbf{E}'_\perp + \sigma_2 \mathbf{B} \times \mathbf{E}'_\perp B^{-1} \tag{3.2}$$

where $\mathbf{E}'_\parallel$ and $\mathbf{E}'_\perp$ are the components of $\mathbf{E}'$ parallel and perpendicular to the magnetic field $\mathbf{B}$, and $\mathbf{E}'$ is the field experienced by the plasma moving with velocity $\mathbf{V}$ so that

$$\mathbf{E}' = \mathbf{E} + \mathbf{V} \times \mathbf{B} \tag{3.3}$$

The usefulness of equation 3.2 in the ionosphere and elsewhere has been discussed by Cowling (1945, 1956), Baker and Martyn (1953), Piddington (1954) and many others. The equation helps formulate physical pictures of current systems and is almost indispensable in some cosmic problems. However, the conductivity components must be related to the true $\mathbf{E}$–$\mathbf{j}$ equation (equation 2.35) or its equivalent, and this is done in the following two sections. Meanwhile we accept equation 3.2 at face value and make some inferences from it.

Particle motions along magnetic field lines are not disturbed by the field so that the current component $\sigma_0 \mathbf{E}'_\parallel$ is independent of the field. Current across the field in the direction of the electric field $\mathbf{E}'_\perp$ is greatly reduced by the magnetic field and is termed the *Pedersen current*. There is also a new *Hall current* ($\sigma_2 \mathbf{E}'_\perp$) perpendicular to the electric field, and this may exceed the Pedersen current. In the case of an isotropically conducting plasma the effect of a boundary is generally to reduce the current flow $\sigma \mathbf{E}'$. For an anisotropically conducting plasma the effects are much more complex and the current in the direction $\mathbf{E}'_\perp$ may be greatly enhanced (Cowling, 1933). Let this electric field lie in the xy plane, with $\mathbf{B}$ along the z axis, then

$$\mathbf{j}_x = \sigma_1 E'_x - \sigma_2 E'_y \qquad j_y = \sigma_1 E'_y + \sigma_2 E'_x \tag{3.4}$$

If there is a boundary perpendicular to the x axis which blocks current flow, then $j_x = 0$ and so $E'_x = (\sigma_2/\sigma_1)E'_y$. Substituting this value in equation 3.4 we have

$$j_y = (\sigma_1 + \sigma_2{}^2/\sigma_1)E'_y = \sigma_3 E'_y \tag{3.5}$$

where $\sigma_3 = \sigma_1 + \sigma_2{}^2/\sigma_1$ is sometimes called the Cowling conductivity. This composite conductivity is most significant in many situations, where it indicates the "effective" conductivity.

The Joule heating of a medium obeying Ohm's Law is $\mathbf{j}.\mathbf{E}' = j^2/\sigma$. In many situations σ_3 replaces σ in the hydromagnetic equations and the Joule heating is given by j^2/σ_3 (Piddington, 1954). This may be shown very simply by adding the current vectors due to an electric field with component E_x perpendicular to **B**. Pedersen current $\sigma_1 E_x$ and Hall current $\sigma_2 E_x$ give a total current $E_x(\sigma_1^2 + \sigma_2^2)^{1/2}$ and a heating rate $\sigma_1 E_x^2 - \sigma_1 j^2(\sigma_1^2 + \sigma_2^2)^{-1}$, or j^2/σ_3. The component of field and current along **B** causes dissipation at a rate j^2/σ_0. In a fully ionized gas we see below that $\sigma_0 = \sigma_3$ and the dissipation is not affected by the field; in a partially ionized gas the field may enhance Joule losses considerably (Section 3.3).

It might be inferred from the discussion given in this section that the main difficulty in plasma theory is in determining the current density from a given electric field. This is not always so, however, because the electric field **E'** itself is often indeterminate. A minor difficulty here is in the correct assignment of axes. In a fully ionized gas and in the approximation $m/M \rightarrow 0$, the axes are such that **V** is the macroscopic velocity of the heavy ions and when $V = 0$, the conductivity tensor is given by

$$\mathbf{j} = [\sigma]\mathbf{E} \qquad (3.6)$$

In the case of a partially ionized gas the choice of axes is more difficult and is discussed in Section 3.3. A much greater difficulty, which is a major problem in the containment of plasma to give con trolled thermonuclear reactions, is the spontaneous growth of electric space-charge distributions. These violate the tendency to neutrality discussed above and limit containment times to about one thousandth of the times needed to provide reactions.

3.2 Fully Ionized Gas

In order to determine conductivity coefficients from the equation of motion (equation 2.31) and the generalized Ohm's law (equation 2.32) we make further simplifications by omitting plasma acceleration and gravitational force and neglecting electron mass. The "force" equation and "conductivity" equation then have the forms

$$\nabla p = \mathbf{j} \times \mathbf{B} \qquad (3.7)$$

$$\mathbf{E} + \mathbf{V} \times \mathbf{B} = (ne)^{-1}\nabla p_i + \sigma_0^{-1}\mathbf{j} \qquad (3.8)$$

Remembering that $\nabla p = \nabla p_i + \nabla p_e$, these may be combined to yield

$$j = \sigma_0\{\mathbf{E} + \mathbf{V} \times \mathbf{B} + (ne)^{-1}\nabla p_e\} - \frac{\sigma_0}{ne}(\mathbf{j} \times \mathbf{B}) \qquad (3.9)$$

which reduces to

$$\mathbf{j} + \omega\tau B^{-1}\mathbf{j} \times \mathbf{B} = \sigma_0\mathbf{E}' \tag{3.10}$$

where

$$\mathbf{E}' = \mathbf{E} + \mathbf{V} \times \mathbf{B} + (ne)^{-1}\nabla p_e \tag{3.11}$$

In the last term of equation 3.9, σ_0 has been replaced by its value given in equation 2.34. If $\mathbf{j}$ is parallel to $\mathbf{B}$ we have $\mathbf{j} = \sigma_0\mathbf{E}'$ as in the case of zero field. If $\mathbf{j}$ is perpendicular to $\mathbf{B}$ we find a result obtained by Cowling (1962):

$$\mathbf{j} = (1 + \omega^2\tau^2)^{-1}\sigma_0(\mathbf{E}' + \omega\tau B^{-1}\mathbf{B} \times \mathbf{E}') \tag{3.12}$$

Thus for both parallel and perpendicular current flow, the relevant electric field ($\mathbf{E}'$) is given by equation 3.11 and so contains an *equivalent electric field* $(ne)^{-1}\nabla p_e$. The relationships between current and electric field $\mathbf{E}'$, together with the σ_0 equation 2.34 yield the three conductivity components.

$$\sigma_0 = \frac{ne^2\tau}{m} \qquad \sigma_1 = \frac{\sigma_0}{(1 + \omega^2\tau^2)} \qquad \sigma_2 = \frac{\sigma_0\omega\tau}{(1 + \omega^2\tau^2)} \tag{3.13}$$

In the solar corona, in a field of 100 gauss, $\omega\tau$ may have a value of 10^6 or more and so σ_1 is reduced below σ_0 by a factor of 10^{12}, while σ_2 is intermediate.

It may seem curious that of the various hydrodynamic forces operative, only the electron pressure gradient contributes as an equivalent electric field. The reason, in the first place, is the convention that when the gas is at rest ($\mathbf{V} = 0$) then only the electrons may move. In a region of pressure gradient Spitzer (1962, Section 2.3) has pointed out that both positive ions and electrons have macroscopic velocities in an inertial system and contribute to $\mathbf{j}$, each proportional to the pressure gradient of the particle type. However, the ion velocity $\mathbf{V}$ is regarded as a mass motion of the medium and contributes to the electric field through equation 3.3; only the electron relative velocity $\mathbf{v}$ provides the current so that

$$\mathbf{j} = -ne\mathbf{v} \tag{3.14}$$

A second reason for the apparent discrepancy is that inertial forces have been neglected. If these are restored, then force equation 3.7 becomes

$$\nabla p + \rho\dot{\mathbf{V}} = \mathbf{j} \times \mathbf{B} \tag{3.15}$$

where $\dot{\mathbf{V}} = \partial \mathbf{V}/\partial t$. Equation 3.9 then becomes

$$\mathbf{j} = \sigma_0 \{ \mathbf{E} + \mathbf{V} \times \mathbf{B} + (ne)^{-1}\, (\nabla p_e + \rho \dot{\mathbf{V}}) \} - \frac{\sigma_0}{ne} (\mathbf{j} \times \mathbf{B}) \qquad (3.16)$$

from which we see that the equivalent electric field is now given by

$$\mathbf{E}' = \mathbf{E} + \mathbf{V} \times \mathbf{B} + (ne)^{-1}\, (\nabla p_e + \rho \mathbf{V}) \qquad (3.17)$$

so that inertial forces also contribute, as one might have expected.

In addition to the current density, a quantity of great importance in plasma theory is the mass velocity $\mathbf{V}_\perp$ of the plasma across the field, which may be found from equations 3.7 and 3.8. Neglecting collisions ($\sigma_0 = \infty$) this is

$$\mathbf{V}_\perp = B^{-2} \{ \mathbf{E} \times \mathbf{B} - (ne)^{-1}\, \nabla p_i \times \mathbf{B} \} \qquad (3.18)$$

which is the quasi-steady solution. When inertial and finite conductivity effects are introduced the equation of motion becomes

$$\mathbf{V}_\perp = B^{-2} [\mathbf{E} \times \mathbf{B} - (ne)^{-1}\, (\nabla p_i + \rho \mathbf{V}_\perp) \times \\ \mathbf{B} - \sigma_0^{-1} (\nabla p + \rho \dot{\mathbf{V}})] \qquad (3.19)$$

The equation of motion of electrons is similar, except that the factor $-(\nabla p_i + \rho \dot{\mathbf{V}})$ is replaced by ∇p_e.

In some cosmic problems, a quantity of major importance is the velocity of the magnetic field lines. Cowling (1962) has shown that by taking the curl of equation 3.9 and using equation 3.14 we obtain

$$\frac{\partial \mathbf{B}}{\partial t} = \nabla \times [(\mathbf{V} + \mathbf{v}) \times \mathbf{B}] + \eta \nabla^2 \mathbf{B} \qquad (3.20)$$

This differs from the hydromagnetic equation 2.5 only by the introduction of the electron peculiar velocity, $\mathbf{v}$, and shows that the field lines tend to move with the electrons rather than with the mass. The difference is usually unimportant, but in the case of magnetic neutral sheets (Section 4.2) it may be very significant.

3.3 Partially Ionized Gases

Most cosmic gases are partially ionized, but some, such as the solar corona and interstellar HII gas, have such small proportions of neutral atoms that their presence may be ignored. Elsewhere, such as in the earth's ionosphere and interstellar HI regions, the effects of neutral atoms are very important and the electrical conductivity is affected more seriously by a magnetic field than that of a binary plasma. The charged particles tend to be transported with the lines of force while the neutral particles are left behind. Collisions between heavy ions and neutral atoms may be severe, lowering the conductivity and unfreezing the plasma from the field lines (ambipolar diffusion). This and other properties of partially ionized gases have been studied by Cowling (1956, 1962), Schlüter (1951), Piddington (1954) and others.

In the partially ionized gases which have been studied in most detail (notably the ionosphere) the neutral atom density is much greater than the ion density. In defining electric fields and currents, and so the conductivity given by the tensor relationship (equation 3.6), the system of axes is set in the neutral-atom gas. Both ions and electrons move in these axes and now the conductivity components all have two terms. Neglecting collisions between ions and electrons, and assuming a single type of ion of mass M, we find

$$\sigma_0 = \sigma_e + \sigma_i = ne^2\tau_e/m + ne^2\tau_i/M \qquad (3.21)$$

$$\sigma_1 = \sigma_e(1 + \omega^2\tau_e^2)^{-1} + \sigma_i(1 + \Omega^2\tau_i^2)^{-1} \qquad (3.22)$$

$$\sigma_2 = \sigma_e\omega\tau_e(1 + \omega^2\tau_e^2)^{-1} - \sigma_i\Omega\tau_i(1 + \Omega^2\tau_i^2)^{-1} \qquad (3.23)$$

Here σ_e and σ_i are the electron and ion contributions to σ_0, τ_e and τ_i are the electron and ion collision periods with neutral atoms, Ω is the ion gyro frequency and the other symbols are defined above.

We will neglect the uninteresting case of a gas in which both ions and electrons collide many times during one gyro period so that the field has little effect. Our first subdivision comprises gases in which electrons may gyrate freely ($\omega\tau_e > 1$) but ions may not ($\Omega\tau_i < 1$). Pedersen

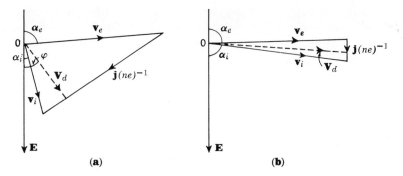

Fig. 3.1. Vector diagrams of ionospheric drifts of electrons (v_e), heavy ions (v_i) and neutral ionization (V_d) relative to the neutral atoms and under the influence of an electric field (E) and magnetic field directed into the paper. (a) In the E-region, about 120 km where $V_d \sim 0.5E/B$. (b) In the lower F region, about 160 km, where $V_d \sim 0.99\ E/B$.

current is then carried mainly by ions and Hall current by electrons, as shown semi-quantitatively in Figure 3.1a which is drawn for iono-spheric E region and discussed in Section 3.5. The alternative situation is when both ions and electrons gyrate and drift freely. Their two Hall drifts more or less cancel so that the Hall current is small, but there remains a significant Pedersen conductivity. This situation is shown by the vector diagram Figure 3.1b, which corresponds to ionospheric F region.

Equations 3.21—3.23 hold for steady electric fields and require modification when the field varies as for hydromagnetic waves. Hines (1953) has derived equations which are applicable for waves of fre-quencies $\omega' \ll \Omega$. These are similar to equations 3.21—3.23 except that τ_e^{-1} and τ_i^{-1} are everywhere replaced by $(\tau_e^{-1} - i\omega')$ and $(\tau_i^{-1} - i\omega')$. They have been used by Karplus et al. (1962) to study the trans mission of hydromagnetic waves through the ionosphere.

When the density of the ion-electron plasma is comparable with that of the neutral-atom gas in a partially ionized gas, equations 3.21—3.23 and also those of Hines are no longer valid. The reason is that the electric field and current of the conductivity equation 3.6 must be defined in a system of axes fixed with respect to the mass of the whole gas. In this system

$$n_i M_i \mathbf{V}_i + n_a M_a \mathbf{V}_a = 0 \qquad (3.24)$$

where the superscripts i and a refer to the ion gas and the neutral-atom gas and the electron mass is neglected. Incorrect choice of axes has caused some confusion on occasions, notably in connection with ionospheric and magnetospheric conductivities. In the lower ionosphere the axes are correctly set in the neutral atom gas and so $\sigma_2 \rightarrow 0$ as the collision frequency tends to zero. At extreme heights, where one might expect σ_2 to approach closer and closer to zero, it rises again to the value ne/B as the system of axes moves from the vanishing neutral-atom gas to the ion gas.

When $n_i M_i$ and $n_a M_a$ are comparable, the medium may be treated as a *fully ionized gas* whose mass motion $\mathbf{V}$ is determined by electro-magnetic forces, together with a *frictional force* due to its motion $(\mathbf{V}-\mathbf{V}')$ relative to the neutral-atom gas whose velocity is $\mathbf{V}'$ (Piddington, 1964). In the momentum equation 2.6, $\mathbf{F}$ is now the frictional term, $\mathbf{j}$ is replaced by $\nabla \times \mathbf{B}$ and the other terms are set at zero, yielding

$$\frac{\partial \mathbf{V}}{\partial t} + (\mathbf{V} - \mathbf{V}') \frac{\zeta}{\tau_a} + \frac{U^2}{B^2} (\mathbf{B} \times \nabla \times \mathbf{B}) = 0 \qquad (3.25)$$

where U is the Alfvén velocity $B(4\pi\rho)^{-1/2}$, ρ and $\zeta\rho$ are the mass densities of the ion-electron gas and the neutral-atom gas, and τ_a is the collision period of an atom. In some situations, such as the ionosphere, the value of τ_a may be much larger than the time scale of the disturbance or wave. This does not invalidate the treatment, but corresponds to the simple case of $V' = 0$. In some situations collisions of atoms with electrons may be more frequent than those with ions; therefore, although the momentum transfer involved is much smaller, they must contribute to reducing τ_a. In spite of these difficulties the Joule loss for hydromagnetic waves of frequency approaching zero is found to agree with that determined by Cowling (1956) by a more rigorous method.

When the neutral-atom gas also moves appreciably its equation of motion is

$$\frac{\partial V'}{\partial t} = (V - V')\tau_a^{-1} \tag{3.26}$$

These two momentum equations may be combined with the field equation 2.5 and solved for disturbances of the form $\exp i(\omega' t - kx)$. Three variations of the hydromagnetic wave may be distinguished:

(1) When $\omega'\tau_a \ll 1$, the two gas components move in unison and the O-type hydromagnetic wave (Section 3.4) has velocity $U(1 + \zeta)^{-1/2}$.

(2) When $\zeta \gg \omega'\tau_a \gg 1$, the waves are electromagnetic waves in a "rigid" medium and are described by well-known skin-effect theory (Watanabe, 1957). The velocity is approximately $U(2\omega'\tau_a)^{1/2}$, so that the medium is now dispersive.

(3) When $\omega'\tau_a \gg 1 + \zeta$, the waves travel as hydromagnetic waves in the ion-electron gas with velocity U. The neutral-atom gas contributes only to absorption.

3.4 Waves in a Magneto-Plasma

An important property of a magneto-plasma is its ability to transmit a variety of waves which carry electric and magnetic perturbations, velocity perturbations and energy, in varying configurations. At first sight it might seem that there is an unlimited variety of possible waves and indeed there may be many waves having slightly different characteristics. However, as shown below, there are four main types of waves, and it is important to understand the physical properties of the plasma which allows each type of wave to propagate. Since these properties are part of the plasma itself, they are discussed in this chapter, while the closely related problem of the generation of the waves is discussed in Chapter 4.

The theory of a pair of *radio waves* in a magneto-plasma was worked out in the early 1920's by Appleton and Hartree who were studying the propagation of radio waves in the ionosphere. A little later a different wave called a plasma oscillation or space-charge electric wave or simply an *electron wave* was studied in the laboratory by Langmuir and Tonks. In the 1940's a *hydromagnetic wave* was discovered and later it was shown by Herlofson and by van de Hulst that there was not only one such wave but three. Finally, if we may resort to such tricks as allowing two or more interpenetrating plasma streams, then any number of waves may be revealed by the plasma dispersion equations. The question is: what is the reality of all these waves and what are their physical natures?

In studying waves in a plasma we use Maxwell's field equations combined with the macroscopic equations of the plasma and seek solutions of the form exp $i(\omega t - \mathbf{k}x)$, where ω is the assumed wave frequency and $\mathbf{k}$ is the wave number vector. These equations may be complex and it is customary to make simplifying assumptions. For example, in the Appleton-Hartree equation (Ratcliffe, 1960) the motions of the (more massive) heavy ions are neglected and for this reason the equations cannot disclose hydromagnetic waves, which depend on these motions. Another curious, but valid, approximation is that the gas is quite cold but at the same time electron collisions are taken into account to explain absorption of the radio waves. This approximation eliminates all trace of electron waves which depend on electron thermal motions. Thus we begin to see why some waves remained undiscovered for so long and what are the connections between wave types and their physical nature.

An accurate wave treatment requires Boltzmann's equation (equation 2.21) and leads to complexities (Stix, 1962; Ginzburg, 1964). While such an approach may be necessary in some problems, an adequate and much simpler method is to combine Maxwell's field equations with Maxwell's equations of momentum transfer (Piddington, 1955). The field equations are given above in Section 2.1, except that displacement current must be added to Maxwell's curl of $\mathbf{B}$ equation 2.2. The momentum transfer equation for the electron gas is

$$m\frac{d\mathbf{v}}{dt} + n^{-1}\nabla p_e + e(\mathbf{E} + \mathbf{v} \times \mathbf{B}) = 0 \qquad (3.27)$$

where $\mathbf{v}$, n and p_e are the electron gas velocity, number density and pressure. The current density $\mathbf{j} = -ne\mathbf{v}$ gives the magnetic perturbation and it becomes a simple matter to find the dispersion equation for weak, plane waves. This equation has three pairs of roots indicating three types of electromagnetic waves. When we put $p_e = 0$, one pair vanishes and we are left with the Appleton-Hartree pair of radio waves (ordinary and extraordinary). The wave made possible by the electron pressure is an electron sound wave. In the absence of a magnetic field this is a purely longitudinal electric wave; it has no magnetic perturbation, since the conduction current **j** is always cancelled by the displacement current.

If we now introduce the additional complication of ion motions, a fourth wave appears called an ion wave. This type of wave may also exist in the absence of a magnetic field and has different characteristics depending on the wavelength being less than or greater than the Debye length. In the former case electric waves similar to the electron waves propagate, while in the latter ordinary sound waves propagate. When a magnetic field is added, the ion wave, like the electron wave, becomes a much more complex electromagnetic wave. In fact it becomes a sort of hydromagnetic wave.

The introduction of ion motions not only allows the ion wave but enormously complicates mathematic formulation of the transverse waves, which have been identified as radio waves. It is desirable, or even necessary, at this stage to divide the spectrum into a radio spectrum and a hydromagnetic spectrum. In the former, ion motion is neglected and we have the O and E (ordinary and extraordinary) radio waves and the electron wave. In the latter, displacement current is neglected and we have O and E hydromagnetic waves which derive directly from the corresponding radio waves. We also have the ion sound wave which has become important, while the electron wave is now unimportant. This gives us a total of four wave types, provided we consider the O and E waves as continuing right through the spectrum. A more detailed discussion of these various wave types is given by Denisse and Delcroix (1963).

In the presence of interpenetrating streams of ions or electrons the dispersion equations develop more roots, apparently revealing new types of waves. These roots reveal real, but usually trivial, phenomena; the perturbation electromagnetic field is impressed individually on each stream and carried along by that stream to appear as a wave. The one exception is the well-known two-stream instability (Bohm and Gross, 1949) which comprises a sort of "electric frictional force" between ion streams with well-separated velocity distributions. This phenomenon is discussed in Section 4.5 as a factor in cosmic radio emissions.

We conclude this section with a brief description of hydromagnetic waves which, as seen in later chapters, are extremely important in many astrophysical situations. Using the equations and symbols of Section 2.1 we will assume perturbations **v** and **b** in the fluid velocity **V** and magnetic field **B**. Since we are not dependent on external forces, pressure gradients or viscous or Joule dissipation, all of these are set at zero and the hydromagnetic equation 2.5 and the equation of motion (equation 2.6) become

$$\frac{\partial \mathbf{b}}{\partial t} = \nabla \times \{\mathbf{v} \times (\mathbf{B} + \mathbf{b})\} \tag{3.28}$$

$$4\pi\rho \frac{d\mathbf{v}}{dt} = (\nabla \times \mathbf{b}) \times (\mathbf{B} + \mathbf{b}) \tag{3.29}$$

If these are now linearized and we note that $\mathbf{B}$ is uniform and $\nabla.\mathbf{b} = \nabla.\mathbf{v} = 0$, then we have

$$\frac{\partial \mathbf{b}}{\partial t} = (\mathbf{B}.\nabla)\mathbf{v} \qquad 4\pi\rho \frac{\partial \mathbf{v}}{\partial t} = (\mathbf{B}.\triangle)\mathbf{b} \tag{3.30}$$

Now take Oz parallel to $\mathbf{B}$ and the equations reduce to

$$\frac{\partial \mathbf{b}}{\partial t} = B \frac{\partial \mathbf{v}}{\partial z} \qquad 4\pi\rho \frac{\partial \mathbf{v}}{\partial t} = B \frac{\partial \mathbf{b}}{\partial z} \tag{3.31}$$

which show that $\mathbf{b}$ and $\mathbf{v}$ are in the same direction and that the magnetic field strength may be increased by shear motion ($\partial v/\partial z$). The equations 3.31 reduce to

$$\frac{\partial^2 b}{\partial t^2} = U^2 \frac{\partial^2 b}{\partial z^2} \qquad \frac{\partial^2 v}{\partial t^2} = U^2 \frac{\partial^2 v}{\partial z^2} \tag{3.32}$$

where $U(\equiv B(4\pi\rho)^{-1/2})$ is the Alfvén velocity, and the velocity of the waves.

In many astrophysical situations macroscopic motions occur with supersonic speeds, and then shock waves are likely to be generated. Such situations develop in the solar atmosphere, near the boundary of the earth's magnetosphere and in interstellar regions and so a theory of shock hydromagnetic waves is required. The basic equations have been given by de Hoffman and Teller (1950) and the structure of these waves has been investigated by Marshall (1955) and Sen (1956) and others.

3.5 The Terrestrial Ionosphere

The origins of ionospheric physics may be traced back for over a century to the suggestion of Gauss in 1839 and Kelvin in 1860 that magnetic daily variations might be caused by an *electrically conducting layer* in the upper atmosphere. This suggestion was adopted by Stewart in 1882 who proposed that convective currents of air established by solar heating moved across the geomagnetic field lines to set up an electric field system and hence electric currents. This "dynamo" theory was later developed by Schuster and Chapman (see Chapman and Bartels, 1940) in quantitative form, and in spite of numerous difficulties was shown by Baker and Martyn (1953) and others to be sound. However, a full and satisfactory quantitative theory is still lacking, mainly because of lack of knowledge of the distribution of electric conductivity.

Meanwhile, following Marconi's transatlantic radio transmission in 1901, the presence of a *reflecting layer* was suggested independently by Kenelly and by Heaviside in 1902. The origin of the layer by solar ultraviolet radiation was suggested by Taylor in 1903, only a few years after Thompson's discovery of the electron. Thus the formation of a plasma sheet above the earth was indicated by two entirely independent experiments: magnetic perturbations and radio propagation. The scientific study of the ionosphere really began with the determination of the height of the reflecting layer by Appleton and Barnett and also by Breit and Tuve, both in 1925. Later the stratified nature of the ionosphere was demonstrated and it was shown that by varying the frequency of the probing radio waves the maximum electron density of each layer could be determined. Intense study of the ionosphere has continued ever since and a wealth of detail has been revealed in the form of complicated patterns of plasma density and electric currents, a complex photochemistry and other electrodynamic effects including hydromagnetic waves driven by the lower atmosphere and others by magnetospheric movements above the ionosphere. These studies have been vitally important in connection with the use of the ionosphere for world-wide radio communications, and reviews have been given by Mitra (1952), Ratcliffe (1960) and Johnson (1965), and in Matsushita and Campbell (1967).

Ionizing radiation of a single wavelength incident on an atmosphere of uniform composition and in hydrostatic equilibrium would create a *Chapman layer*. At great heights there is low ion density because of the low gas density; at low enough levels there is little ionization because most of the radiation has been absorbed. Intermediate, there is a well-defined maximum N_m of electron and ion density. This has *critical* radio frequency f given by $N_m = 1.24 \times 10^4 f^2$ electrons/cm^3; f is expressed in megacycles per second and is the lowest frequency which can penetrate the layer. During the daytime, several distinct "layers" or regions are recognized, although the separation between them is not distinct. In order of increasing altitude and increasing ion concentration these are the D, E, F_1 and F_2 regions. The first three are really ledges on the lower slope of the F_2 region, except sometimes at night when E develops into a separate peak. Above the F_2 peak the electron density decreases monotonically out to several earth radii, merging into the "protono-sphere" or "magnetosphere," which is discussed in later sections. Until relatively recently there had been few studies of the magneto-sphere, which was considered an extension of the ionosphere (if it was considered at all). Now it might be more appropriate to consider the ionosphere as the base of the much more extensive magnetosphere. During the day the E region (90–140 km) is ionized by soft solar x rays to an electron density of about $10^5/cm^3$. The F_2 region (200–1300 km) is ionized by the HeII 304 Å line and perhaps others to an electron density of about $5 \times 10^5/cm^3$ at sunspot minimum and about $2 \times 10^6/cm^3$ at sunspot maximum. All of these densities correspond to latitudes between about 30° and 40°.

Electrodynamic effects in the ionosphere are numerous and often complex, but the most important may be listed according to their driving mechanisms as follows:

(1) First we have an *atmospheric wind* which results from solar heating and gives rise to the atmospheric dynamo mentioned above. The air moves with velocity V, mainly horizontally, across the magnetic field B_0 and so induces an electric field $V \times B_0$. This field drives an electric current J which may be defined as the current per unit strip of the ionosphere integrated through the full depth of the ionosphere. Because of spatially varying conductivity, space-charge electric fields develop to give a total electric field $E + V \times B$ and we have the relationships

$$\nabla . J = 0 \qquad J = [\Sigma] (E + V \times B_0) \qquad (3.33)$$

where $[\Sigma]$ is the integrated conductivity tensor. Equations 3.33 are based on the simplifying assumptions that the ionosphere may be regarded as a thin conducting sheet with no current flowing upwards or downwards. These assumptions are not strictly true, but are not un-reasonable approximations and do enormously simplify a very difficult problem. The conductivity components may be found from equations 3.21—3.23 and a suitable model of the ionosphere.

Wind and dynamo theory has been reviewed by Maeda and Kato (1966). The winds, and the resulting electrodynamic effects, are dominat-ed by a diurnal component, but there is a semi-diurnal component and also a prevailing wind, constant in direction (mainly westward) and magnitude over long periods, though changing seasonally.

(2) Other ionospheric fields and currents are driven by convective

motions within the magnetosphere, which, in turn, are driven by the *solar wind*. Such effects are most varied and complex and since they originate far above the ionosphere are discussed in Chapters 8 and 9.

(3) Finally, there are a number of *small-scale motions* driven by atmospheric waves and turbulence, by meteors, by instability of large-scale ionization drift motions and by other forces. These motions may be studied by radio-reflection methods provided they involve irregularities in the density of the ion-electron gas. Studies of such irregularities are then directed to their origins and to their subsequent drift motions under the influence of electrodynamic and other forces. Such studies were initiated by Martin (1953) and reviewed by Kato (1965).

We may start by assuming a uniform, stationary, slightly ionized gas subject to an electric field $\mathbf{E}$ and a magnetic field $\mathbf{B}_0$ into the paper in Figure 3.1. The velocity vectors of the electron gas $\mathbf{v}_e$ and ion gas $\mathbf{v}_i$ are obtained from the corresponding currents transported by these gas components (since $\mathbf{j}_e = -ne\mathbf{v}_e$ and $\mathbf{j}_i = ne\mathbf{v}_i$). The Pedersen components (direction $\mathbf{E}$) of these two currents are given by equation 3.22 after multiplying by E; the Hall components (direction $\mathbf{B} \times \mathbf{E}$) are given by equation 3.23. Substituting from equation 3.21 we find the two components of electron drift with similar expressions for ion drifts.

$$v_e(\text{Pedersen}) = \frac{-e\tau_e E}{m(1 + \omega^2\tau_e^2)}$$

$$v_e(\text{Hall}) = \frac{-e\omega\tau_e^2 E}{m(1 + \omega^2\tau_e^2)} \tag{3.34}$$

The angle a_e between the vectors $\mathbf{v}_e$ and $-\mathbf{E}$ (Figure 3.1) is given by $\tan a_e = \omega\tau_e$; similarly the direction of $\mathbf{v}_i$ is given by $\tan a_i = \Omega\tau_i$. The vector diagram may now be completed by the modified current vector $\mathbf{j}(ne)^{-1}$, which has two components provided by the drifts $\mathbf{v}_e$ and $\mathbf{v}_i$. The other components of $\mathbf{v}_e$ and $\mathbf{v}_i$ are identically equal to the mass drift $\mathbf{V}_d$ of the ion-electron gas as a whole.

These relationships allow us in a very simple manner to determine the motions of the various components of the ionospheric plasma. Above 90 km the electron gas moves with virtually the full Hall drift E/B_0 in a direction $a_e = \pi/2$. On the other hand, the ion gas moves nearly parallel with $\mathbf{E}$ at 90 km ($a_i = 0°.02$) and swings steadily to the direction $\mathbf{E} \times \mathbf{B}$ at greater heights ($a_i = 80°$ at 160 km). The two diagrams of Figure 3.1 refer to levels near 100 km and near 160 km, and we note that in the former the Hall current and the Pedersen mass drift are substantial, while in the latter only Pedersen current and Hall drift are important.

From such diagrams we may also determine how ion-electron gas tends to accumulate in clouds or irregularities. This is revealed by the equation of density change

$$\frac{\partial n}{\partial t} = q - \epsilon n - \nabla.(n\mathbf{V}_d) \qquad (3.35)$$

where q and ϵ are the rates of electron formation and decay. In an initially homogeneous medium, irregularities may form as a result of divergence of $\mathbf{V}_d$, and this may result from the introduction of electric fields which have divergence as in the electron "pump" described below.

A difficulty arises in Figure 3.1 when the medium is not homogeneous: we have a drift $\mathbf{V}_d$ of electron-ion gas with a component along $\mathbf{E}$, and yet electrons drift almost exactly along $\mathbf{E} \times \mathbf{B}$. The answer is that the theory is based on the assumption of approximate electrical

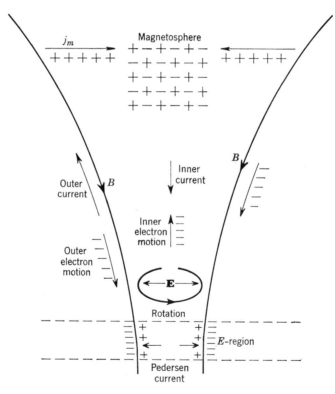

Fig. 3.2. An electron-ion "pump" comprising a magnetic force tube BB connecting the ionospheric E-region to the magnetosphere. When the tube rotates clockwise as shown, a radial electric field E develops which drives a current system as shown. Neutral ionization (equal quantities of ions and electrons) is pumped out of the tube at the E-region level and perhaps into the tube at magnetospheric levels.

neutrality so that wherever ions congregate the more mobile electrons must also congregate. They may do so by flowing up or down field lines to or from a reservoir of electrons in the magnetosphere. In Figure 3.2 an electron "pump" is shown in cross section, comprising a magnetic force tube BB which dips into the ionosphere and rotates as indicated, being driven from above. The motion $\mathbf{V}$ of the field lines through the ionosphere induces an electric field $\mathbf{E} = -\mathbf{V} \times \mathbf{B}$, radially outwards from the axis of the force tube. In the low E region, Hall current (carried by electrons) flows in circular paths around the axis and causes no charge accumulation. Throughout the ionosphere, Pedersen current (carried by ions) flows radially and causes an accumulation of positive ions around the surface of the tube. These cause electric space charge which must be neutralized, and so electrons flow down the surface field lines as shown. The radial flow of Pedersen current also causes an accumulation of negative space charge near the axis of the tube, and this causes electrons to flow upwards. The net results are accumulations of ion-electron gas near the surface of the tube in the lower E region and near the axis of the tube in the distant magnetosphere. The vertical currents shown in Figure 3.2 are parts of the dynamics of the tube of force, which also transmits a torque.

The drift motions of irregularities are the subject of a very extensive literature and here we may only refer briefly to one or two important aspects. One is that the conductivity within a cloud differs from that outside, and this changes the electric field and drift pattern in and around the cloud. The result is that the cloud does not, as one might expect, drift with the velocity $\mathbf{V}_d$ determined above. Instead it has a velocity

$$\mathbf{V}_c = \{\mathbf{j} \times \mathbf{B}_o - (\sigma_2/\sigma_1)B_o\mathbf{j}_\perp\} \, (nM/\tau_i + nm/\tau_e)^{-1} \qquad (3.36)$$

where σ_1 and σ_2 are again the conductivity components, and $\mathbf{j}_\perp$ is the current component orthogonal to $\mathbf{B}_o$. It is easy to identify the first term on the right with the velocity $\mathbf{V}_d$, because it gives a balance between the electromagnetic force $\mathbf{j} \times \mathbf{B}_o$ and the collisional frictional force for the ion-electron gas. The second term is due to the localized drift: ions and electrons leave one side of the cloud and accumulate on the other side to add a wave motion to the mass-drift motion. Near the 100 km level the wave velocity may be the larger by a factor of about ten.

Finally, we must remember that plasma may move along the magnetic field lines under the influence of gravity or pressure gradients. For example, plasma near the equator may lie above the F_2 region, but because of the downward slope of the field lines away from the equator it slides down and away. The result is two maxima of ionization on either side of the equator, or the F_2 equatorial anomaly (Duncan, 1960).

References

Baker, W. G. and Martyn, D. F., 1953, *Phil. Trans. Roy. Soc.* A **246**, 281.
Bohm, D. and Gross, E. P., 1949, *Phys. Rev.* **75**, 1851 and 1864.
Chapman S. and Bartels J., 1940, *Geomagnetism,* Clarendon Press, Oxford.
Cowling, T. G., 1933, *Mon. Not. Roy. Astron Soc.* **93**, 90.
Cowling, T. G., 1945, *Proc. Roy. Soc.* A **173** 453.
Cowling T. G. 1956 *Mon. Not. Rot. Astron. Soc.* **116** 114.
Cowling T. G. 1962 *Rep. Progr. Phys.* **25** 244.
de Hoffman F. and Teller, E., 1950, *Phys. Rev.* **80**, 692.
Denisse, J. F., and Delcroix, J. L., 1963, *Plasma Waves,* Interscience Publishers, New York.
Duncan, R. A., 1960, *J. Atmos. Terr. Phys.* **14**, 89.
Ginzburg, V. L., 1964, *The Propagation of Electromagnetic Waves in Plasmas,* Pergamon Press, Oxford.
Hines, C. O., 1953, *Camb. Phil. Soc.* **49**, 299.
Johnson, F. S., ed., 1965, *Satellite Environment Handbook,* 2nd Ed., Stanford University Press, Stanford.
Karplus, R., Francis, W. E. and Dragt, A. J., 1962, *Planet. Space Sci.* **9**, 771.
Kato, S., 1965, *Space Sci. Rev.* **4**, 223.
Maeda, K. and Kato, S., 1966, *Space Sci. Rev.* **5**, 57.
Marshall, W., 1955, *Proc. Roy. Soc.* A **233**, 367.
Martyn, D. F., 1953, *Phil. Trans. Roy. Soc.* A **246**, 306.
Matsushita, S. and Campbell, W. H., eds., 1967, *Physics of Geomagnetic Phenomena,* Academic Press, New York.
Mitra, S. K., 1952, *The Upper Atmosphere,* Asiatic Society, Calcutta.
Piddington, J. H., 1954, *Mon. Not. Roy. Astron. Soc.* **114**, 638 and 651.
Piddington, J. H. 1955 *Phil .Mag.* **46** 1037.
Piddington J. H., 1964, *Space Sci. Rev.* **3**, 724.
Ratcliffe, J. A., 1960, *Physics of the Upper Atmosphere,* Academic Press, New York.
Schlüter, A., 1951, *Z. Naturforsch.* A **6**, 73.
Sen, H. K., 1956, *Phys. Rev.* **102**, 5.
Spitzer, L., 1962, *Physics of Fully Ionized Gases,* Interscience Publishers, New York.
Stix, T. H., 1962, *The Theory of Plasm Waves,* McGraw-Hill, New York.
Watanabe, J., 1957, *Sci. Rep. Tohoku Univ. Ser.* 5 **9**, 81.

CHAPTER 4

Electrodynamic Effects of Universal Occurrence

In later chapters various electrodynamic phenomena are discussed in relation to their cosmic settings: the earth's environs, the sun and the solar system, stars and the interstellar medium, and galaxies and the intergalactic medium. Some of these phenomena recur in different parts of the universe on different scales and it is convenient to discuss them first in a general way rather than in relation to their regions of occurrence. This applies particularly for effects whose mechanisms are not understood, such as magnetic field merging and annihilation in neutral sheets. It is fairly certain that this phenomenon is important in connection with solar flares, the growth and decay of the geomagnetic tail and elsewhere, but it is not understood. It is desirable, therefore, to discuss possible basic mechanisms before considering these in relation to their settings. On the other hand, some phenomena are fairly well understood and so are best discussed in relation to their cosmic surroundings.

Good examples of the latter type of effect are the various magnetohydrodynamic instabilities. These have been mentioned briefly in Section 2.2 and references given for their further study. One instability which needs special mention because of its widespread occurrence is the *Rayleigh-Taylor instability*. In its simple hydrodynamic form this concerns two fluids of different densities superposed one over the other in a gravitational field or, alternatively, accelerated towards each other. If the heavier fluid is on top of the lighter fluid, then the interface is unstable and the heavier fluid drips down through the lighter one to a lower energy configuration. The electrodynamic version of this instability occurs when a plasma is supported by a magnetic field, perhaps aided by a hotter, less dense plasma. If a slight ripple travels across the interface, bending the field lines, then the denser plasma flows along the field lines to the troughs where it is further compressed, while the less dense plasma rises. This instability is found in solar

43

magnetic fields (Section 5.1), supernova shells (Section 11.5) and near the surfaces of galactic disks (Section 11.1).

The most important of the variety of cosmic electrodynamic phenomena is the creation of magnetic flux or induction and of magnetic energy. Since they depend on magnetic fields, all other cosmic electrodynamic phenomena, and perhaps even the present components of the universe, are secondary to the magnetic fields. A closely related problem, and one which should be discussed within the same context, is the maintenance of magnetic fields in a steady state for periods much greater than those of natural decay by Ohmic dissipation. These two problems are discussed in Section 4.1 together with a third, which is the origin of a "fossil" magnetic field from which all other fields developed. The problem of *magnetic amplification* is relatively simple and so is discussed in greatest detail in connection with galaxies where it is most clearly important (Chapter 12). The problem of *cosmic dynamos* is much more difficult and much of the work on dynamo theory is directed towards explaining the earth's magnetic field. For these reasons this theory is reviewed in somewhat greater detail in this chapter.

Perhaps second in importance only to mechanisms of magnetic amplification are mechanisms of *field annihilation* which are discussed in Section 4.2. When two slabs of plasma with oppositely directed fields are placed in contact the field lines merge and convert their energy to particle kinetic energy. This effect is extremely important not only in the solar system but also in determining magnetic configurations in galaxies, including radio galaxies, and perhaps in connection with cosmic-ray acceleration. Most current theories appear suspect because they treat the problem as one of fluid dynamics, yet at the same time depend on the existence of boundary layers smaller than the ion gyro radius.

A phenomenon of wide occurrence and great importance is that of *magnetic interchange motions* in which one magnetic force tube with its plasma changes place with another tube, more or less parallel to the first. In the laboratory such motions occur as the *interchange instability* (or flute or ripple instability) and similar phenomena will occur in nature, perhaps within the earth's magnetosphere. There are, however, other forms of interchange motions which are not results of instabilities, but are driven by forces of external origin applied across a boundary or along the magnetic force tube. One of these is the magnetosphere interchange motions driven by frictional interaction with the solar wind. (Section 8.5) Such motions are discussed in Section 4.3.

In many parts of the universe particles are observed with energies far above the average of the ambient gas. There appears to be a number of mechanisms by which a select few particles are accelerated and these are discussed in Section 4.4.

Finally, we briefly review the half dozen or so *radio-emission mechanisms*, which have revealed so much more of the universe during the past two decades. It should be borne in mind that these phenomena are quite distinct from all others discussed in this chapter, and also from emissions of light, infrared and x radiation, in that they represent a relatively small transfer of power, which has no major influence on the dynamical history of the emitting object. The significance of the radio emissions lies in the extreme sensitivity of the detecting devices and the fact that some of the emission mechanisms are so specialized that they reveal a great deal about conditions at the source.

4.1 Origins of Cosmic Magnetic Fields

It seems that the universe is almost completely permeated by magnetic fields, extending from the terrestrial field to fields in intergalactic space. Notable exceptions are planets such as Venus (Section 10.4), which lack a field of internal origin, and also are both electrically conducting and are immersed in a medium whose magnetic field reverses periodically. Other regions of zero field, very limited in space and time, are magnetic neutral sheets discussed in the following section. Naturally, the first problem concerning cosmic magnetic fields is their origin.

Attempts have been made to explain cosmic fields as a fundamental property of massive rotating bodies and as a result of thermoelectric currents, but these theories have met with no success. It is now generally accepted that the fields of planets, stars and galaxies originated from "fossil" fields, developed and maintained by "dynamo" action. The theory of fossil magnetization asserts that the magnetic field at present existing in a body is a relic of the field with which it was born. The time of free decay of a magnetic field is given by equation 2.12 and in the case of the earth this time is only about 10^4 years so that some mechanism for its maintenance is required. In the case of the sun and magnetic stars, the time is of order 10^{10} years which is much more than the age of most stars, so that the problem of maintenance is removed. However, it is replaced by another problem, that the configuration of the fields is not easily explained in terms of fossil fields unless these have been greatly altered, again by dynamo action. The same considerations apply to galactic magnetic fields, including those of radio galaxies.

A star or a planet is likely to be born with a substantial field, because of the interstellar magnetic field known to exist (Section 11.1). Indeed, the difficulty here is that the field tends to be too large because the formation of a star from average interstellar gas requires a reduction in linear dimensions by a factor of about 10^8. This will cause an increase in the strength of a frozen-in field by a factor of about 10^{16}, from about 3×10^{-6} gauss to 3×10^{10} gauss, which is excessive. This difficulty is discussed in Section 11.2, but meanwhile it is clear that an adequate stellar fossil field is available.

This is not the case for galactic fields, because there is no direct evidence for a universal magnetic field. However, a strong argument in its favor is the apparent impossibility of explaining galactic fields in any other way. Just as a long time is required for a field to diffuse out from a large body, so a long time is required for a field to diffuse into the body. For a galaxy, this is much longer than the lifetime of the universe and so we must, apparently, accept the existence of an intergalactic magnetic field. An extension of this argument is given in Chapters 12 and 13 where a hydromagnetic theory of galactic forms and activity is described. Such a theory, which is based on the assumption of a large-scale intergalactic field, seems capable of explaining many characteristics of galaxies including the outbursts which lead to the formation of radio galaxies, the variety of forms of "normal" galaxies and the peculiar features of the central regions of spiral systems. In the absence of satisfactory alternative explanations this theory provides strong evidence for an intergalactic field.

If we accept the existence of an intergalactic field, then the origin of the "seed" field is removed one step further but not solved. In the laboratory a magnetic field may be created by connecting a wire to a voltaic cell, but in the universal plasma we lack adequate electrochemical energy. An alternative energy source which may be worth considering is that involving nuclear processes such as the decay of unstable isotopes. A requirement is the alignment of the spin axes of many particles, in which case non-conservation of parity would ensure a macroscopic electric current carried by electrons ejected in a preferred direction (Piddington, 1962).

Once we have a seed field, no matter how weak, this may be built up by an amplification process, which is based on the fact that motion of material across the lines of force of an existing field generates currents. These may have a configuration so that they amplify the original field and continually change its topology. Starting with the generalized Ohm's Law (equation 2.35) and multiplying both sides by .j we have

$$\mathbf{E}.\mathbf{j} + (ne)^{-1}\nabla p_e.\mathbf{j} = \frac{j^2}{\sigma_0} + \mathbf{V}.(\mathbf{j} \times \mathbf{B}) \tag{4.1}$$

On the left-hand side is the work done by the electric field driving the current and also by the electron pressure gradient driving the current. On the right-hand side is the energy dissipation due to collisions and a term which is the product of the Lorentz force $\mathbf{j} \times \mathbf{B}$ and the mass motion $\mathbf{V}$. When $\mathbf{V}.(\mathbf{j} \times \mathbf{B}) > 0$, work is done by the electromagnetic field on the plasma, accelerating it or moving it against pressure gradients; the magnetic field energy decreases. When $\mathbf{V}.(\mathbf{j} \times \mathbf{B}) < 0$, the non-magnetic forces of equation 2.6 drive the plasma against the magnetic resistance and so pump energy into the magnetic field. This is *magnetic amplification* and, in more complex configurations, the *self-excited dynamo*. A criterion for amplification is that the dynamo term be numerically greater than the Joule loss term, and if we replace $\mathbf{j}$ by the field gradient $B/4\pi L$, where L is a characteristic length, then the criterion is

$$4\pi\sigma_0 VL > 1 \tag{4.2}$$

A simple mechanism for magnetic field amplification is shown in Figure 4.1 in which plasma rotates anticlockwise about an axis perpendicular to the paper and more rapidly in the center. This winds the field into an ever-tightening spiral and continuously increases the field strength. There is no steady-state condition available in this simple topology and so it is not a dynamo. However, the continuously evolving field leads to many interesting effects which may have application in galaxies and, in particular, radio galaxies. These effects are discussed in Chapters 12 and 13.

The reasons why simple field structures such as that of Figure 4.1 are inadequate as dynamos were first investigated by Cowling (1933). In the first place the continued winding up steadily reduces the distance between the neutral sheet turns and so reduces L in equation 4.2. Also, as magnetic stresses increase, these brake the plasma motion and reduce V until the amplification criterion is no longer satisfied. Finally, increasing magnetic pressure along the rotational axis must be taken into account. These effects are of great interest in connection with galaxies but they cannot account for the steady maintenance of the earth's surface dipole field. If this is to be maintained by dynamo action then the internal field must be considerably more complicated.

Dynamo theory is complex and the approach has been mainly

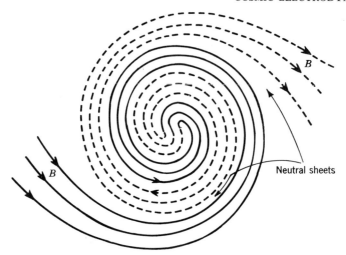

Fig. 4.1. Magnetic field lines *B*, which were originally straight, are wound into a double spiral under the influence of differential rotation of the plasma. The central region has rotated anticlockwise about one revolution relative to the outer region. Full lines and dashed lines represent oppositely directed fields separated by a spiral "neutral sheet."

kinematic. According to Cowling (1965), the following conditions are essential for dynamo generation and maintenance of large-scale fields such as that of the earth and dipole-type fields of stars.

(1) The mass motions that supply the energy must have a regular, though not necessarily steady, pattern. Although there has been some disagreement, it does not seem possible to power a dynamo by turbulent motion and the only models which appear plausible all involve laminar motions with well-defined patterns.

(2) At least two types of regular motion must be present, one to stretch the existing field lines and the other to transform the new magnetic flux into a new state. This is found in the well-developed geomagnetic model of Bullard and Gellman (1954), where a toroidal field is first generated from a poloidal field by non-uniform rotation. Four stages of this development are shown schematically in Figure 4.2, where faster internal (easterly) rotation draws a poloidal field line into two circular loops. The second type of motion is that of convective cells giving rising and falling motions near the equator. This is an essentially non-axially symmetric motion and generates a new poloidal field from the toroidal. The resulting field resembles a dipole field outside the globe, but has a large toroidal component in its interior. The dynamo motions themselves were postulated and the forces needed

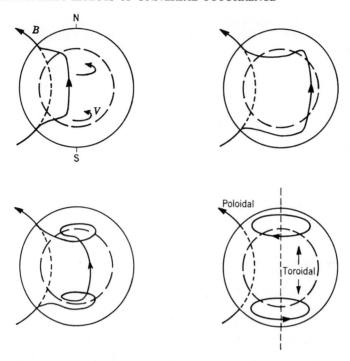

Fig. 4.2. Successive stages in the development of a toroidal magnetic field from a poloidal field in a differentially rotating fluid sphere.

to maintain them were not considered in detail but only shown to be of the same order of magnitude as those likely to be met in the earth's core.

The double-motion dynamo was demonstrated more rigorously by Backus (1958) and by Herzenberg (1958) who took as an adequate, if unrealistic, model a large solid conducting sphere at rest, in which were imbedded two small conducting spheres which could rotate about particular axes. With sufficiently rapid rotation a steady field was possible for about half of the possible orientations of the two axes of rotation.

(3) The third requirement of a dynamo is a suitable mechanism for destroying magnetic flux. The motions that generate the wanted field will simultaneously amplify other components such as the toroidal field of Figure 4.2. When these other components have served their purpose they must be destroyed or they will accumulate until they block the mechanism. The theories described above all rely explicitly on finite electrical conductivity which permits the lines of force to diffuse through the material.

In the case of stars the conductivity is so great that there is difficulty in bringing freshly generated field from the interior to the exterior. In this case, as opposed to that of the earth, magnetic flux may emerge by "tearing" of the surface plasma. Arched magnetic field lines thrust above the surface and the material lifted with them then slides down the field lines to uncover new surface plasma (Section 5.1). This overcomes the difficulty that if magnetic lines of force are sufficiently well frozen into a star, they are unable to emerge through the surface and build up an external field. Some dynamo theories are concerned with the solar cycle and so are discussed in more detail in Chapter 5. Further reference is made to stellar magnetic fields and their growth in Chapter 11.

4.2   Magnetic Field Annihilation and Reconnection

When two plasma slabs having oppositely directed magnetic fields are in contact, the boundary between them is called a *magnetic neutral sheet*. In order that the magnetic field **B** should reverse, a current must flow in the boundary layer given by the curl of **B** (equation 2.2.) Also, since the magnetic pressure at the neutral sheet is zero, the two fields and their plasma must be held apart by plasma pressure alone. Thus the plasma pressure in the neutral sheet must exceed that inside the slabs by an amount equal to the magnetic pressure in the slabs. For this reason the neutral sheet is also called a *current pinch*. When the fields do merge they neutralize or annihilate one another and release magnetic energy which is transformed to particle kinetic energy; it is this effect in particular which has attracted attention to the phenomenon.

The possible significance of neutral lines and sheets as sources of energy and electric fields has been discussed by Giovanelli (1946, 1948), Dungey (1953), Sweet (1958) and Parker (1963). The phenomenon has been of interest mainly in connection with solar flares (Section 5.4) and the difficulty has been to explain the conversion of a large amount of magnetic energy (more than 10^{32} erg) in times of about 10^3 sec. A mechanism for accelerating some of the plasma particles to cosmic-ray energies is also sought in the neutral sheet.

In its simplest form the problem reduces to the rate of diffusion of a magnetic field out of each of a pair of plasma slabs of dimensions L and conductivity σ. As seen in Section 2.2, the characteristic time is $4\pi\sigma L^2$. In the solar corona L must be at least 10^4 km and σ about 10^{-5} emu, and the time becomes more than 10^6 years. The discrepancy, as far as flares are concerned, is by a factor of 10^{10} or more and would seem to exclude the neutral sheet as a useful energy source.

If the neutral sheet is placed in the chromosphere where the gas is only partially ionized, the conductivity is reduced somewhat (Gold and Hoyle, 1960). However, as Parker (1963) has shown, the enhancement falls far short of that required. Here we must refer to the discussion of Section 3.3 of the conductivity of partially ionized gases. It is customary in calculating the conductivity of such gases to use axes fixed in the neutral-atom gas, and this is valid for most purposes in the lower ionosphere. However, when the mass densities of the ion-electron gas and the neutral-atom gas are comparable this method is not valid. In the limit, when the physical effects (through collisions) of the neutral-atom gas are small, it plays no significant part in the determination of the conductivity and yet the calculated conductivity is enormously reduced because of the incorrect choice of axes.

A big step towards an adequate field annihilation rate is made when account is taken of flow of plasma along the field lines. Sweet's (1958) mechanism for enhanced diffusion is illustrated in Figure 4.3, where

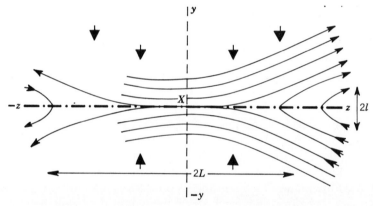

Fig. 4.3. A magnetic neutral sheet or current pinch separating two plasma regions ($y > 0, y < 0$) in which the magnetic fields are oppositely directed. Field lines diffuse together near the line vertically from the paper through X (sometimes referred to as a neutral line or neutral point) and then contract in directions $\pm z$.

two regions of conducting fluid containing approximately uniform magnetic fields are pressed together as shown by the full arrows. Above the boundary $y = 0$ the field is in the z direction and below the boundary it is in the $-z$ direction. As soon as the regions are pressed together, fluid will begin to flow along the field lines: this squeezing out of the fluid allows the two fields to approach more and more closely to the boundary. Thus the field gradient near the boundary increases above that in the static model, as do also the diffusion and annihilation rates.

An order-of-magnitude calculation of the diffusion rate more clearly illustrates the physical principles involved. In Figure 4.3 field lines reconnect at a neutral line or region near X and the reconnected lines then contract away from the plane of symmetry, $z = 0$, with velocity V, ejecting fluid from the ends of the pinch region. To replace this loss, fluid moves with velocity v towards the $y = 0$ plane from either side. The length of the neutral sheet is $2L$ and the thickness of the transition layer $2l$. If we neglect any compression effects, then conservation of mass gives $vL = Vl$. Conservation of energy and momentum balance combine to show that fluid is ejected with velocity approximately equal to the Alfvén velocity $U = B(4\pi\rho)^{-1/2}$, where B is the field strength away from the transition region. Finally, the velocity with which the fields are merging is given approximately by $v = (\sigma l)^{-1}$. Combining these three equations we have

$$v = \left(\frac{U}{\sigma L}\right)^{1/2} \qquad\qquad l = \left(\frac{L}{\sigma U}\right)^{1/2} \qquad (4.3)$$

The time required to dissipate a pair of magnetic fields of length and thickness both equal to L is then L/v or $(\sigma L^3/U)^{1/2}$. In the solar corona, in a region where the electron density is $10^{10}/\mathrm{cm}^3$, field strength 100 gauss, dimensions 10^4 km and conductivity 10^{-5} emu, the dissipation time is about three months, which is still too large by several orders of magnitude. Nevertheless a large reduction (from 10^6 years) has been effected.

Following Parker's analysis, attempts have been made to remove this discrepancy by introducing more complicated and efficient patterns of plasma flow. One promising model is that of Furth et al. (1963), which invokes the "tearing-mode" instability. In Figure 4.3 merging occurs only near the plane $z = 0$ and the reconnected field lines must contract a distance L before they eject all of their fluid out from between the two merging regions. In the tearing mode, merging occurs at reduced intervals along the z axis and the reconnected field lines contract into complete loops. This allows accelerated merging to take place between these loops. The pinched current in Figure 4.3. must flow into the paper, and in the tearing mode this current sheet is torn into strips lying parallel to the x axis. Other hydromagnetic flow models have been proposed by Petschek (1964), Coppi (1965), Sturrock and Coppi (1966) and others and reviewed by Akasofu (1966). Generally speaking these models retain the main features of the Parker-Sweet model, in that they are two-dimensional and that the actual diffusion rate is controlled by the electrical conductivity, which means physically that it depends on the dynamically rather insignificant effect of electron collisions. The main efforts have been directed towards increasing the diffusion rate by decreasing the diffusion dimension l above. In a way this approach has been so successful, in some models, that it has defeated its own purpose by reducing l to a value much less than the mean collision distance and even less than the ion gyro radius. The former limits the applicability of the hydromagnetic method and the latter renders it, and the model, invalid.

The apparent failure to provide an adequate theory of magnetic field annihilation and reconnection results from lack of experimental data on the one hand, and from the very limited assumptions made in the models so far developed. The former difficulty is being remedied by systematic observations of solar flares and their associated plasma motions and other effects. (Section 5.4), and by observations made in the neutral sheet of the earth's magnetic "tail" (Chapters 7 and 8).

The limitations of the theoretical approaches may be listed as follows:

(1) Relative motion of plasma and magnetic field lines is attributed solely to particle collisions, according to the equation of diffusion (equation 2.12) of field lines through a stationary medium. In the solar corona, for a typical ion and electron density of $10^9/cm^3$ and temperature 10^6K, an ion travels a distance of about 3 km before colliding with an electron. If we put $L = 3 \times 10^5$ cm and $\sigma = 10^{-5}$ emu in the diffusion equation, then the characteristic time is about 10^7 sec, which is again too large by a factor of about 10^5. This may indicate the inefficiency of the diffusion mechanism in the solar corona.

(2) A two-dimensional structure is adopted.

(3) A nearly time-independent solution is sought. The neutral sheet problem is one of the most challenging in the field of cosmic electrodynamics, or indeed, in the whole field of astrophysics.

Models should be considered which do not suffer from the above limitations.

In Chapter 3 we derived two equations which describe the motion of plasma in the presence of a magnetic field. These are based on some approximations, but they do offer some variety of effects which cause motion of field lines across the plasma. The first is the equation of mass motion (equation 3.19) which includes the collision–diffusion term in σ_0; we consider that in the solar atmosphere and in other regions containing neutral sheets this term in unimportant. Another force which contributes to $V_\perp$ is pressure gradient, and the term in ∇p_i may be large. However, this gradient must be perpendicular to the neutral sheet and so the plasma drift is along the isobars and ineffective. We are left with the terms in $\mathbf{V}$ and $\mathbf{E}$, but before discussing these we should refer to the second equation of motion (equation 3.20). This shows that when the electron gas moves relative to the ion gas (that is, when a current flows), the magnetic field lines follow the electron gas rather than the plasma mass.

The **E** term in equation 3.19 allows the plasma to move across the field lines with the Hall drift. This could clearly be of importance in helping field merging, provided a space-charged field could be applied and maintained. However, the development of a model requires a knowledge of the prior history of the plasma and fields as well as the full geometry of the surroundings. This line of attack appears very difficult, but may prove rewarding.

Alternatively, an examination of the effects associated with the term $\dot{\mathbf{V}}$ in equation 3.19 suggests that this may provide a breakthrough in neutral-sheet theory. In terms of the model shown in Figure 4.3, all of these effects result from the passage of a *compression wave* in the direction x (into the paper). The acceleration is then $\dot{V}_x$.

(1) We consider only the component of $\dot{V}_x$ caused by the electromagnetic field, so that from the force equation 3.15 we have

$$\rho \dot{V}_x = j_y B \tag{4.4}$$

with j_y away from the neutral sheet on both sides. The current j_y is part of the hydromagnetic compression wave which we have introduced by applying a piston in the plane of the paper of Figure 4.3. It will be seen that we have removed all three limitations of the earlier theories: the new term may be very much larger than the σ_0 term, and the model is now three-dimensional and time-dependent.

(2) Through Maxwell's curl of **B** equation (equation 2.2) the current j_y causes a gradient in magnetic field strength and pressure in the direction $-x$. This gradient is the same on both sides of the neutral sheet, and is also a part of the compression wave thought responsible for accelerated merging.

(3) The drift term in equation 3.19 due to $\dot{V}_x$ is away from the neutral sheet on both sides. The two plasma slabs with frozen-in fields are already moving together with appropriate velocities as shown by the arrows of Figure 4.3. An opposite drift is now imposed on the plasma of each slab, but not on the field lines which merge accordingly. The magnitude of the opposing drift is $\rho \dot{V}_x / enB$.

(4) The current j_y is carried by electrons which move, together with magnetic field lines, into the neutral sheet. Because of their greater mobility the electrons are disposed of by moving along the field lines. The protons are ejected in the direction x and it is found that if they acquire most of the energy of the destroyed field, then their motion suffices to carry the whole neutral-sheet current.

This model of neutral-sheet merging has been discussed briefly (Piddington, 1968) and is referred to again in Chapters 7 and 8. It is clearly very complex and the above discussion inadequate, because it uses the hydromagnetic method on a boundary layer which turns out to have a thickness of one gyro radius (at the velocity of ejection).

4.3 Magnetic Interchange Motions

Consider an incompressible fluid in which the magnetic field is everywhere parallel to the z axis, but not everywhere uniform. Fluid motions in the xy plane carry the field lines as rigid rods, and through any fluid element the magnetic intensity $\mathbf{B}$ is constant in magnitude and direction. Such motions can exchange regions where B is strong for regions where B is weak, some lines of force moving in one direction and others in the opposite direction. These motions are called *interchange* motions and they are important in many astrophysical situations as well as in the laboratory.

In the above simple situation there is no interchange of energy between the field and the fluid and no change of potential energy of the to system or from kinetic energy; the system is in neutral equilibrium. Now suppose that a perturbation electric current flows parallel to $\mathbf{B}$ and as a result $\mathbf{B}$ rotates in the yz plane as x increases. The directions of the lines of force in the yz plane are now different for different values of x and in general interchange motions are forbidden because skew field lines are in the way. Thus a shear magnetic field tends to have greater stability than one without shear by inhibiting some interchange motions or, alternatively, by deforming the field and changing the magnetic energy.

It should not be thought, however, that all interchange motions are repressed in plasmas with sheared or bent field lines. In the above example all field lines lie in planes perpendicular to the x axis and motions parallel to these planes are possible. In general, any motions are possible for which shear is confined to surfaces containing magnetic field lines. An important example is seen in Ferraro's law of isorotation: in a differentially rotating body of plasma such as a star the condition for equilibrium is that angular velocity is constant over a surface generated by rotation of a line of magnetic force about the axis. As we shall see, this theorem has interesting extensions when applied to conducting bodies rotating as solids and having fully conducting atmospheres (stars) and partially insulating atmospheres (the earth).

So far interchange motions have been considered only from a kinematic point of view and it is now necessary to consider forces which drive and retard the motions, and also the interchange of energy forms involved. In cosmic situations the forces involved are diverse and it is convenient to consider them under three headings: internal forces, external retarding forces and external driving forces.

Internal Forces

The various forces in a magneto-plasma must balance according to the equation of motion (equation 2.6). Expanding the Lorentz force term according to equation 2.11 this may be written

$$\nabla p + \rho\dot{\mathbf{V}} - \rho\mathbf{g} = \frac{1}{4\pi}(\mathbf{B}.\nabla)\mathbf{B} - \nabla\left(\frac{B^2}{8\pi}\right) \qquad (4.5)$$

It is generally possible to simplify this equation by neglecting one or two terms on the left-hand side. If inertial effects are not important, then it reduces to the magnetostatic equation 2.14 which corresponds to an equilibrium situation. This equilibrium may be stable or unstable against interchange motions and its nature may be determined by assuming perturbations in the form of weak waves of the form exp $i(kx - \omega t)$. The method is essentially similar to that used in the analysis of waves propagating in the magneto-plasma. However, in the equilibrium problem, the coefficients in the differential equations of the perturbed quantities are functions of position and the analysis is much more difficult. Some of the modes revealed may increase exponentially instead of oscillating, in which case the equilibrium is unstable. References to such analyses are given in Chapter 2.

A very much simpler method of determining instabilities is to consider the total potential energy W of a plasma and its internal and external field, which is

$$W = \int\left\{\frac{B^2}{8\pi} + \frac{3p}{2} + \rho\phi\right\} d\tau \qquad (4.6)$$

where ϕ is the gravitational potential ($\nabla\phi = -\mathbf{g}$), $d\tau$ is a volume element and the integral extends over the volume occupied by the plasma and field. If, for the time being, we neglect Joule losses and viscosity, then the system is nondissipative and the total energy, given by the sum of W and the kinetic energy, is constant. Let the system be perturbed by imposing an arbitrary displacement and calculate the change δW in W. In an equilibrium system δW is zero to the first order of the perturbations and it is necessary to include second-order terms. If δW is then found to be negative, the system is unstable and interchange motions will occur spontaneously and develop under the influence of internal forces.

Perhaps the most interesting cases of cosmic interchange motions are found in the magnetospheres of the earth and Jupiter. At first sight it would seem that the law of isorotation would prohibit such motions, as it would in the case of a rigidly rotating, fully conducting body such as a star. However, the existence of a layer of non-conducting gas between the solid earth and the magnetosphere changes the situation completely and allows a class of motions to occur in which the magnetic field intensity at each point remains unchanged, but the plasma convects (Gold, 1959). These motions may be understood by reference to Figure 4.4, where two magnetic force tubes X and Y, having equal magnetic

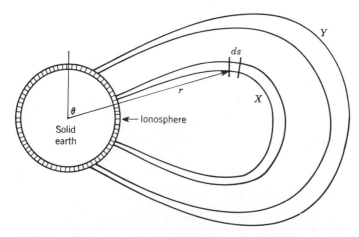

Fig. 4.4. Two magnetic force tubes, X and Y, of equal flux may interchange in the earth's magnetosphere because the ionosphere and solid earth are separated by an insulating layer of gas which in effect "cuts" the field lines.

fluxes, exchange places. In the absence of an insulating lower atmosphere shown as the dashed ring, any displacement or rotation of the tubes would involve bending or twisting of the field lines between the magnetosphere and solid earth into which they are frozen. This would require currents flowing in the lower atmosphere and since, in fact, these are forbidden there can be no resistance to motions of the tubes caused by the presence of the solid earth. The two tubes and their plasma content will move so as to change places provided their total energy $W_X + W_Y$ thereby decreases.

In some magnetospheric problems the total magnetic energy remains constant. On the other hand, a new term must be introduced to take account of centrifugal force, which is sometimes important. The total potential energy associated with a force tube of unit magnetic flux is then (Sonnerup and Laird, 1963)

$$W_{X,Y} = \int \left(\frac{\gamma p}{\gamma-1} + \rho\phi - \frac{\rho}{2}\omega^2 r^2 \sin^2\theta \right) \frac{ds}{B} \qquad (4.7)$$

The integral is along the length of the tube whose area is B^{-1} and the first term represents the sum of the internal energy and the confinement energy (or work done against the magnetic field in order to create space for the gas), p and γ being the gas pressure and ratio of specific heats. The last term is a pseudopotential energy corresponding to the centrifugal force due to rotation ω at axial distance $r \sin \theta$ (see Figure 4.4).

Gold, and Sonnerup and Laird, investigated interchange stability in the earth's magnetosphere for isothermal and adiabatic interchanges. Gold concluded that if the particle density decreased faster than $r^{-20/3}$, the distribution would be unstable, and that interchange could take place slowly for a density decrease greater than r^{-4}. Melrose (1967) concluded that Jupiter's magnetosphere would be unstable beyond a radial distance of six planetary radii for a decrease in density faster than r^{-4}.

External Retarding Forces

A factor which has been neglected in the above discussion is the viscous drag of the feet of force tubes X and Y as they move through the much more dense (compared with the magnetosphere) ionosphere. Apart from tending to inhibit interchange motions, the other effects of this viscous drag are of great importance. These are the ionospheric electric currents responsible for some geomagnetic disturbances and ionospheric motions seen in auroral forms and elsewhere. They are discussed in Chapters 8 and 9.

A more detailed analysis of magnetospheric interchange stability has been given by Chang et al. (1965). They used a Liouville equation in place of the magnetohydrodynamic method and individual particle motions are based on the descriptions given by Northrup and Teller (see Section 8.4). They also introduce the damping effect of the ionosphere, and conclude that this probably stabilizes the Van Allen radiation belts of energetic particles, but only during the day, when the ionosphere has a high conductivity. Swift (1967) has considered growth rates of the interchange motions taking into account all of the above factors.

External Driving Forces

The possibility of externally driven interchange motions was suggested by Piddington (1960) and extended to continuous motions by Axford and Hines (1961). Chapman and Ferraro had shown earlier that the earth's magnetic field would be compressed and contained on the solar side by pressure of the solar wind. The extension to interchange motions involved some sort of "electromagnetic friction" tangential to the geomagnetic cavity which is discussed in more detail in Chapters 7 and 8. Also envisaged was a geomagnetic tail into which fresh field lines were carried by the solar wind and from which other field lines returned to the closed magnetosphere; these motions also comprise a complicated class of interchange motions.

In order to describe the *vortex pattern* of interchange motions within the closed magnetosphere we refer again to Figure 3.2. This is highly schematic, in that the top of the magnetic force tube is supposed to lie in the equatorial plane while the bottom dips into the ionosphere at high latitudes. In the figure shown, the corresponding bend in the tube is removed. External forces, originating in the solar wind, introduce a torque and an anticlockwise (looking down the tube from the top) rotation of the top section of the tube. A hydromagnetic twist wave travels down the tube with Alfvén velocity and the field lines become helixes. For the present purposes we may neglect the internal hydrodynamic effects and consider only the torque of the solar wind and the viscous drag of the ionosphere.

To facilitate quantitative discussion, the further simplification is made that the magnetic force tube is cylindrical in form. A system of axes r, θ, z is chosen with the initial unperturbed field B_z. A perturbation field b_θ is introduced at the top of the cylinder and travels downward with velocity $B_z(4\pi\rho)^{-1/2}$. At a given instant the wave front has moved a distance z_0 and at this surface the disturbance is terminated. Current $j_z(r)$ flows down the field lines to the surface $z = z_0$, where a current sheet $J_r(r)$ (integrated through the wave front) provides termination. Maxwell's equation for the curl of $\mathbf{B}$, together with $\nabla \cdot \mathbf{j} = 0$, give

$$\frac{1}{r}\frac{\partial}{\partial r}(rb_\theta) = 4\pi j_z \qquad \frac{1}{r}\frac{\partial}{\partial r}(rJ_r) - j_z = 0 \qquad (4.8)$$

From these, $b_\theta = 4\pi J_r$ and in the ionosphere this current (Pedersen) provides a Lorentz force $J_r B_z$ per unit increment of r, integrated over the height of the ionosphere. It is this force which limits the rate of interchange motion. It also provides positive space charge at the tube boundary, to create an upward current there which closes the circuit and places spatial limits on the electromagnetic disturbance.

The above model acts as a plasma pump, as shown in Section 3.3. The radially outward electric field drives Hall current (J_θ) clockwise in the ionosphere and, as seen in Chapters 8 and 9, this current is responsible for some polar magnetic disturbances. Finally, the possible effects of externally driven interchange motions in Jupiter's magnetosphere are discussed in Section 10.2.

4.4 Acceleration of Charged Particles

The acceleration of charged particles is the basis of cosmic electrodynamics; the acceleration may be that of a plasma as a whole, or individual particles may be favored. It is surprising how widespread is the latter effect, and it seems that wherever there is magnetohydrodynamic activity a small proportion of the particles attains energies far beyond the average. In a solar flare, for example, mass plasma motions provide average proton energies of a few keV, yet some protons attain energies of 10^9 eV or more which is larger by a factor of about 10^6. In the earth's magnetic field a small proportion of the trapped protons attains energies beyond 10^6 eV, again about 10^6 times greater than the particle energy of the ambient plasma. As we shall see in the following chapters, the same phenomenon is found in Jupiter's vicinity, in various stellar and interstellar objects, notably the remarkable Crab Nebula, and above all in the central systems of many galaxies.

In discussing this phenomenon it is desirable to consider the motions of individual particles in relation to an electromagnetic field configuration, and also perhaps of an ambient plasma. The latter must be taken into account in connection with time-varying magnetic fields, which are limited to accord with possible plasma motions. Particle acceleration is given by equation 2.1, with momentum **p** replacing $m\mathbf{v}$ so as to take account of possible relativistic effects.

$$\dot{\mathbf{p}} = e(\mathbf{E} + \mathbf{v} \times \mathbf{B}) \tag{4.9}$$

Since a time-stationary magnetic field cannot change the particle energy, it is clear that such a change requires either an electric field or a time-dependent magnetic field. Such fields may be divided more specifically into *electric space-charge fields*, and magnetic fields whose *lines of force are moving* in the system in which acceleration takes place. We will first discuss acceleration in the absence of electric space charge.

Betatron Acceleration

Betatron acceleration was first proposed by Swann (1933) and is an application of Faraday's law, which states that in the presence of a time-dependent magnetic field, an electromagnetic force is induced (in a conductor, if present). In a closed loop, where dl is an element of arc length, bounding an open surface A through which there is a field B, the induced electric field satisfies the equation

$$\int_l \mathbf{E}.dl = -\frac{d}{dt}\int_A B.dA \qquad (4.10)$$

which is obtainable by integrating the curl of $\mathbf{E}$ equation (equation 2.3) and using Stokes' theorem. The particle is accelerated as long as the magnetic flux through the area A continues to increase.

A simple, quantitative evaluation of betatron acceleration is available by use of a particle's magnetic moment $\mu = mv_\perp^2/2B$, where $v_\perp$ is the component of velocity perpendicular to the field $\mathbf{B}$. As we shall see in Section 8.4, this quantity is constant for all field changes except those which occur within a time comparable with the particle gyro-period. Thus if B increases, the perpendicular kinetic energy component must increase proportionally.

An example of this process might be the general compression of the geomagnetic field caused by an increase in solar wind pressure. If the magnetic field is stationary but the particles are convected by an electric field and magnetic field gradient into a region of stronger magnetic field, a similar acceleration will take place. This process can also be looked on as a change in energy as the particle moves to a region of different electrical potential. The mechanism has also been invoked by Swann and others to explain the acceleration of some flare particles, but here it failed by overlooking the effect of the plasma into which the field is frozen. An increase in particle energy by a factor of 10^6 requires a corresponding increase in field strength and also of plasma density, which is quite out of the question.

The Fermi Mechanism

The Fermi mechanism of particle acceleration stems from Fermi's (1954) far-reaching concept of particle reflection at a magnetic mirror. The gross effect can be considered in its simplest case as a reflection of the particle from a moving mirror. If the mirror moves toward the approaching particle, the particle gains energy in the collision similar to a ball bouncing from a moving tennis racket. Microscopically the process results from the electric field induced by the moving mirror. In the reference frame of the plasma, whose motion carries the mirror, there is no E field. Hence, in the stationary frame a field exists, $E = -\,V \times B$, which accelerates the particle to higher velocity. An example of such a mirror is shown in Figure 4.5; a proton approaches from the

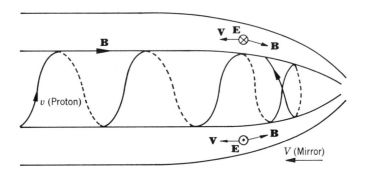

Fig. 4.5. A magnetic "mirror" and a proton in process of reflection and accelera-tion. The magnetic field **B** increases in strength towards the right and the plasma into which the field is frozen moves (**V**) towards the left.

left and the mirror moves to the left. In the region where **B** has com-ponents perpendicular to **V**, an electric field is experienced by the proton which is into the paper at the top of the diagram, and out from the paper at the bottom. This field accelerates the particle and, in combination with the magnetic acceleration e/m (**v** × **B**), ejects it at a higher velocity along the field lines. It is not clear from this discussion which component of particle velocity is increased. However, after reflection the particle finds itself in a region where the field strength is the same as it was before reflection and since its magnetic moment, and hence its trans-verse velocity, must be unchanged the acceleration must be along the field lines. The mechanism of Figure 4.5 shows the basic equivalence between the betatron and the Fermi acceleration processes, although in the former transverse velocity $v_\perp$ is increased and in the latter longi-tudinal velocity $v_\parallel$ is increased.

Further insight into the Fermi mechanism is provided by assuming a second magnetic mirror joined to the one shown in Figure 4.5. This one will be to the left, and moving towards the right with velocity V, so that the mirrors approach one another with velocity $2V$. If the mirror separation is L then it is possible to demonstrate that $v_\parallel L$ is constant, from the fact that at each reflection the particle velocity increases by $2V$. The number of reflections per second is $v_{\parallel}/L$ so that

$$\frac{dv_{\parallel}}{dt} = \frac{2v_{\parallel}V}{L} = -\frac{v_{\parallel}}{L}\frac{dL}{dt} \qquad (4.11)$$

which on integration gives $v_{\parallel}L = $ constant. In the more general case with a complex field configuration and continuously varying $v_{\parallel}$, the result is stated as $\oint v_{\parallel}ds = $ constant and is termed the "longitudinal invariant." If proof of this invariance is accepted (Section 8.4), then Fermi acceleration is easily demonstrated with the above argument inverted.

In random magnetic fields, acceleration by the Fermi mechanism is very slow because a particle encounters almost as many receding mirrors as approaching mirrors and only experiences a statistical acceleration. In the magnetic bottle model every encounter provides acceleration and the process is much more rapid (first-order Fermi acceleration). Even in this model, however, the process may be slow. To accelerate a particle from hydromagnetic to relativistic velocities requires about 100 reflections and if the distance scale is 10^5 km (as in the solar atmosphere), then the time taken is minutes or hours.

Magnetic acceleration mechanisms have been discussed by Parker (1958) and fully reviewed by Hayakawa et al. (1964). They have been considered in relation to the origin of galactic cosmic rays by Ginzburg and Syrovatskii (1964).

Electric Space-charge Fields

Acceleration of charged particles by electric space-charge fields is simple in concept; the difficulty is mainly in the detection of such fields and so in developing theories of their origin. Such fields may be directed across or along the magnetic field lines and in the former case they tend to be ineffective because the particle merely drifts perpendicular to **E** and **B** and gains no energy. There are many references to the literature to large electric potential fields in the earth's magnetosphere and to particle acceleration by these fields, even though they are orthogonal to **B**. This is correct, but may be regarded in either of two ways. One way is to consider the magnetic field lines at rest (valid if the field configuration is time-independent), with ambient plasma convecting with velocity **V**. The induced field **V** × **B** must then be cancelled by a reverse space-charge field and energetic particles drifting across this field are accelerated by this **E** field. An equally valid approach is to consider the magnetic field lines convecting with the plasma (interchange motions) so that there is no space-charge field. The fast particles are then accelerated by the moving magnetic field as for betatron acceleration.

It would seem then, that to be fully effective space-charge fields must lie along the field lines, and in this case they must be very weak as a result of the very high electrical conductivity of the ambient plasma. It is possible that exceptions to this rule occur in magnetic neutral sheets or as a result of streaming instability (Swift, 1967), but the mechanisms are not fully understood. However, even very weak waves directed along the field lines may be effective if they extend far enough or are applied for a long enough period. The latter possibility occurs for electric space-charge waves travelling along the magnetic field-lines. A particle trapped in the potential trough of such a wave is carried along at the phase velocity of the wave. If this velocity increases at a sufficiently slow rate the particle is accelerated with the wave, because as it tries to escape it experiences the electric field of the wave and is accelerated continuously. This mechanism may be important in the earth's magnetosphere, the solar atmosphere and elsewhere.

Electromagnetic Waves

Finally, we consider particle acceleration by electromagnetic waves, both radio waves or hydromagnetic waves. The most commonly considered case is that of resonance between the particle cyclotron motion with angular frequency Ω, and the electric perturbation of a circularly polarized radio wave moving along the field line. It is not necessary for the wave frequency to be Ω, because particles will be selected having velocity $v_\parallel$ along the field line such that

$$\omega \pm k_\parallel v_\parallel = \Omega \tag{4.12}$$

where ω and $k_\parallel$ are the wave frequency and component of the wave number along the field line. The wave selects groups of particles (from a spectrum of particle velocities) travelling in either direction so that $v_\parallel$ and Ω are matched to provide particle acceleration. Conversely, other groups of particles also close to resonance are able to give up their energy to the wave so that the mechanism may generate packets of fast particles from waves or waves from packets.

Another acceleration mechanism, which seems to be important in the magnetosphere, is by resonance with hydromagnetic waves. As we have seen, Fermi acceleration occurs by conservation of the longitudinal invariant when two magnetic mirrors move towards one another. Acceleration also occurs if one or both mirrors oscillate so that they are approaching whenever the particle is being reflected (Section 8.4).

4.5 Cosmic Radio Emission Mechanisms

In the optical region, cosmic electromagnetic radiation comes mainly from bound-bound transitions between discrete atomic or molecular states, free-bound transitions during recombination, and free-free transitions in the continuum. Some components of cosmic radio emission are generated in an analogous way. The hydrogen line emission at 21 cm, the OH emission and a number of other lines result from bound-bound transitions, while thermal radio emission from interstellar and coronal gas is due to free-free transitions or bremsstrahlung. In addition there are a number of other mechanisms which are responsible for the major part of cosmic radio radiation. We might ask why radio frequencies are so favored.

In a vacuum a charged particle radiates only while it is undergoing acceleration and, neglecting relativistic effects for the time being, the power radiated is proportional to $e^2 \mathbf{v}^2$, where $\mathbf{v}$ is the velocity. The very important point is that the emission is proportional to the square of the charge, and this is independent of relativistic effects. A second important relationship results from the fact that a wave of frequency f arises mainly from accelerations which vary appreciably in time $1/f$. In this time an electron (say) moves a distance v/f and so the emission arises from accelerations which vary during motions through distances of about v/f. These two relationships and the fact that f is much smaller for radio than for light waves answer the above question in the following ways.

In the first place the distance scale, or wavelength, $\lambda \sim v/f$ is some 10^5 times or more larger for radio waves, and this allows an individual electron to be accelerated effectively by fields which provide radio but not light emission. One of these is the cosmic *magnetic field* which allows cyclotron emission for non-relativistic electrons, and synchrotron emission for relativistic electrons. It should be noted that synchrotron *optical* radiation is quite possible and is observed in a few regions, notably the Crab Nebula (Section 11.5). However, the electron energy needed is large, typically 10^{12} eV or more, and this phenomenon is not common. Another type of field which is capable of providing radio, but not optical, emission is the macroscopic or larger-scale *electric field* due to a cloud of electrons or an electric space-charge wave.

In the second place, the single electron whose radiation we have been considering may also be replaced by a bunch of electrons which are then accelerated in unison in either of the above large-scale fields. This is possible because of the large permissible dimensions of the bunch, and does not apply for optical emission. A cloud of n electrons, of dimensions much less than v/f and moving in unison, will radiate n^2 times the power of a single electron, and therefore each electron radiates n times the power it would radiate if alone. This *stimulated emission* occurs whatever the accelerating mechanism and raises the interesting possibility of stimulated synchrotron emission. The size of the electron bunch must be considerably smaller than the gyro radius because each electron emits only in a narrow cone.

Interpretation of cosmic radio emissions requires an understanding not only of the emission processes but also of absorption, refraction and polarization. The former is simply the reverse of the thermal, cyclotron or synchrotron emission processes. It imposes a limit on the *brightness temperature*, or temperature of a black body which gives the same emission, equal to the thermal energy or temperature of the radiating electrons. This is not the case for stimulated emission by a bunch of particles, where the energy of the whole bunch must be considered.

Understanding of refraction is aided by the concept of refractive index or ratio of wavelengths in free space and in the medium concerned. In the absence of relativistic effects, the refractive index depends essentially on the "plasma frequency" f_0 and the gyro or cyclotron frequency f_B, and these are given by

$$f_0{}^2 = \frac{Ne^2c^2}{\pi m} \qquad f_B = \frac{Be}{2\pi m} \qquad (4.13)$$

where as before N, e, m are the electron number density, charge (emu) and mass. For quasi-longitudinal propagation (more or less along the magnetic field), the refractive index is given by

$$u^2 = 1 - f_0{}^2 \, (f^2 \pm ff_B)^{-1} \qquad (4.14)$$

revealing two waves, an ordinary (O) wave for the upper sign and an extraordinary (E) wave for the lower sign. In the case of the O wave, the refractive index is real when $f^2 + ff_B > f_0{}^2$ (pass band), but becomes imaginary for lower frequencies (stop band, evanescent waves). In the case of the extraordinary wave the behavior is more complex but it is easily seen that a wave which is generated in a region where $(f_0{}^2/f^2 + f_B/f) > 1$ (u is imaginary) is unable to escape into regions where f_0 and f_B are small. This restriction is most important in the theory of radiation from the solar corona and Jupiter's ionosphere because emission mechanisms which are theoretically attractive meet this escape difficulty.

We will now discuss the physical processes involved in the different emission mechanisms (other than line emission).

Thermal Radio Emission

Thermal radio emission from hot plasma depends on random thermal velocities of the electrons being changed at collisions. The absorption coefficient is given by

$$\kappa = \frac{\nu f_o^2}{c f^2} = \frac{\zeta N^2}{f^2 T^{3/2}} \tag{4.15}$$

where ν is the collision frequency, T is the temperature and ζ is a slowly varying function whose value in most regions of interest may be taken as 0.12. After travelling a distance S through this medium the wave retains a fraction $e^{-\kappa S}$ of its original power and so has lost a fraction $(1 - e^{-\kappa S})$. The quantity κS is the optical depth of the sheet of thickness S and at a temperature T it may, by Kirchhoff's law, radiate with brightness temperature T_b and brightness b, given by

$$T_b = T(1 - e^{-\kappa S}) \qquad b = \frac{2kT}{\lambda^2}(1 - e^{-\kappa S}) \tag{4.16}$$

where k is Boltzmann's constant. When $\kappa S \gg 1$ the sheet is opaque and $b \propto \lambda^{-2}$; when $\kappa S \ll 1$ the sheet is optically thin and

$$b = \frac{2\zeta k N^2 S}{c^2 T^{1/2}} \tag{4.17}$$

A thermal emission spectrum is shown as curve (a) in Figure 4.6, the dashed extension being black body emission, while at short wavelengths the brightness becomes independent of λ when the sheet or cloud becomes transparent.

Gyro or Cyclotron Radiation

Gyro or cyclotron radiation is emitted from an electron gyrating in a magnetic field at the frequency f_B given above. The emission and the corresponding absorption processes are extremely efficient, and they correspond to the E wave. However, thermal gyro radiation cannot escape from the plasma because of the stop band $f < f_B + f_o^2/f$ which must be met as the radiation travels to regions of decreasing field strength and plasma density.

The interest in gyro radiation concerns electrons which are not only gyrating but are moving through the plasma with velocities which give their radiation a substantial Doppler shift. Such a shift may increase the frequency emitted above that of the E-wave resonance frequency of the surrounding medium and so allow the wave to escape. Theories based on emission from bunches of electrons moving along magnetic

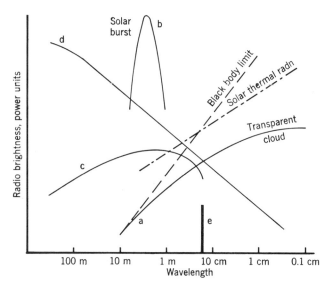

Fig. 4.6. Some typical radio spectra of cosmic sources. Curve a: Thermal emission from a cloud which has become optically thin at wavelengths below about 1 cm. The dashed extension is for an optically thick cloud or black body. Curve b: Typical cyclotron emission, and also a spectrum of plasma oscillations. Curve c: Synchrotron emission from an electron of energy about two million volts in a field of one gauss (this curve is really an envelope of the numerous harmonics). Curve d: Synchrotron emission from a typical cosmic-ray distribution. Curve e: The 21 cm hydrogen line emission.

field lines in the solar corona and in Jupiter's magnetosphere have been developed by Ellis (1965) and others to explain observed emissions. The type of emission spectrum given is a narrow band of frequencies near the cyclotron frequency with a displacement and spread caused by Doppler shift. Such a spectrum is shown as a curve b Figure 4.6.

Synchrotron Emission

Synchrotron emission, or magnetic bremsstrahlung, was first proposed as a source of cosmic radio emission by Alfvén and Herlofson and by Kiepenheuer (1950). The theory of this phenomenon has been reviewed by Ginzburg and Syrovatskii (1965, 1969) and others.

An electron gyrating in a magnetic field radiates the gyro frequency f_B, but as its speed increases it also radiates harmonics. At speed $3c/4$ (energy 2.6×10^5 eV) the second harmonic is as strong as the fundamental. When the kinetic energy greatly exceeds the rest mass energy it radiates almost entirely in the harmonics, the frequency of maximum emission being

$$f_m = \frac{1}{2} f_B \left(\frac{E}{mc^2} \right)^2 \tag{4.18}$$

where E is the total energy. The envelope of the emission spectrum of a single electron is shown as curve c in Figure 4.6. The power radiated is given by

$$P = 3.8 \times 10^{-6} \; B_\perp^2 E^2 \; \text{erg/sec} \tag{4.19}$$

This allows us to find the lifetime of the particle.

$$T \sim 10^{-2} \; B_\perp^{-2} \; E^{-1} \; \text{years} \tag{4.20}$$

Emission from an assembly of cosmic rays of typical power spectrum has a radio spectrum shown as curve d of Figure 4.6. The synchrotron process, like thermal emission, acts in reverse to give self absorption and to impose a black body limit on the source brightness depending on the electron "temperature" or energy E. This effect is most important in quasi-stellar sources, as it imposes a lower limit on the size and upper limit on the magnetic field strength.

Plasma Oscillations

Plasma oscillations may occur in a simple non-magnetic plasma when the particles have thermal motions. They are electron sound waves transmitted by the thermal motions at velocity $V = v(1 - f_o^2/f^2)^{-1/2}$, where v is roughly the electron thermal velocity. They have no associated magnetic field because the convection electric current is cancelled by a displacement current, and they cannot propagate outside the plasma itself. When a magnetic field is present they become more complex, but their general properties are not changed and one would expect them to be of no interest to the radio astronomer.

Nevertheless, they have two properties which do make them of considerable interest. In the first place they will grow spontaneously in a region where two electron streams interpenetrate at a relative speed greatly in excess of any thermal speeds. Consider a perfectly homogeneous medium comprising two such streams, together with enough positive ions to provide electrical neutrality everywhere. This system is in unstable equilibrium as seen by introducing a perturbation in the form of a limited region of slight excess electron density. The excess electrons belong to the two streams and the region is called a space-charge cloud. The electric field of the cloud slightly slows electrons approaching in the two streams and accelerates them on leaving.

The net result is that all passing electrons spend more time in the cloud than they would if it had no excess charge. They add to the excess charge and cause even more slowing down of the next sections of the streams and so cause a continuous growth of the space charge. Such growing waves or clouds are investigated mathematically by formulating appropriate dispersion equations and seeking imaginary roots. Care must be taken, however, in the identification of an instability among the much more common phenomenon of a *reflected wave* which it may resemble in the dispersion equation (Buneman and also Sturrock, in Drummond, 1961).

The second interesting characteristic of plasma waves (or moving electron or ion clouds) is that the radio emission provided by an accelerating field is proportional to the square of the total number of electrons in the cloud. Such clouds, accelerated in electromagnetic fields of sufficiently large scale, provide coherent or stimulated emission. (Ginzburg and Ozernoi, 1966).

Coherent or Stimulated Emission

As we have seen, emission beyond the black body limit is not possible where electrons suffer random acceleration. On the other hand some cosmic sources have brightness temperatures approaching or exceeding 10^{20}K, and so far above the electron temperature. Most notable of such sources are the pulsars (Section 11.6), some quasars (Section 13.2), the local source in the Crab Nebula (Section 11.5), Jupiter and Io (Sections 10.2 and 10.3), the sun (type II and other type bursts, discussed in Section 5.6), some flare stars (Section 11.4) and some galactic OH clouds (Section 11.4).

The explanation is coherent emission or stimulated emission by large numbers of electrons moving in unison. The best known example of coherent emission is a radio or television transmitting antenna where many cold electrons move in an orderly manner to radiate powerfully. Since the electron cloud must greatly exceed atomic dimensions, free-free emission may not be enhanced in this way. However, coherent cyclotron and coherent Cerenkov radiation (see below) are possible, and plasma waves or oscillations (see above) which interact with plasma inhomogeneities to provide radio emission (Ginzburg and ZheleznNikov, 1958) are clearly examples of a coherent mechanism.

In the case of synchrotron radiation, it would appear from intuitive arguments that a small bunch of relativistic electrons moving in a magnetic field in a vacuum could provide coherent emission (or experience negative reabsorption). However, this simple mechanism is clearly ineffective because the electron bunch would immediately scatter under the influence of mutually repulsive electric forces. On the other hand, if the vacuum is replaced by a suitable plasma, then a suitable choice of energy spectrum of the relativistic electrons may allow coherent synchrotron emission (McCray, 1966; Zheleznyakov, 1966). Zheleznyakov (1967) has applied this theory to explain the very high radio brightness temperatures of quasars, the Crab local source, *UV* Ceti type flare stars and type IV solar bursts.

Cerenkov Radiation

Cerenkov radiation results from the passage of a charged particle (or cloud of particles) through a medium with velocity above the wave velocity in the medium. There are four wave modes in a magneto-ionic medium (Section 3.4) and any of these may be stimulated. These are the electron and ion sound waves and the transverse electromagnetic waves (radio at high frequencies, hydromagnetic at low frequencies). In the case of the radio waves it is the *E* wave which has a low enough velocity to allow it to be generated by a Cerenkov process. However, as mentioned above, it cannot escape from the source region because of a stop band. Nevertheless Cerenkov emission may be important because the *E* wave, or a plasma wave or even a slow hydromagnetic wave, may interact with plasma irregularities to create other wave modes including the radio O mode which can escape.

Other Emission Mechanisms

Finally, we briefly mention atomic and molecular line emission which has allowed us to measure the distribution of neutral hydrogen and hydroxyl molecules in our galaxy. The 21 cm line is shown as curve e in Figure 4.6. Doppler shifts of this line allow accurate velocity distributions of the gas to be made thereby adding greatly to the power of this method of investigation.

When a radio photon collides with a fast electron it may gain energy and move to a different part of the radio spectrum (or even to higher spectral regions). In a way this inverse Compton process may be regarded as a radio emission mechanism, and so requires mention here. In fact it may be important in determining the power output and spectra of some synchrotron sources and is mentioned specifically in connection with the Crab Nebula (Section 11.5).

An important property of cosmic radio waves is their polarization or configuration of their electric and magnetic vectors. This property provides evidence of the emission mechanism and sometimes of the nature of the intervening cosmic plasma. An adequate discussion is too lengthy for inclusion here, but has been given by Gardner and Whiteoak (1966). Other aspects of wave propagation have been discussed in Section 3.4 and in references given there.

References

Akasofu, S.-I., 1966, *Space Sci. Rev.* **6**, 21.

Axford, W. C. and Hines, C. D., 1961, *Canad. J. Phys.* **39**, 1433.

Backus, G., 1958, *Ann. Phys. (New York)* **4**, 372.

Bullard, E. C. and Gellman, H., 1954, *Phil. Trans. Roy. Soc.* A **247**, 213.

Chang, D. B., Pearlstein, L. D. and Rosenbluth, M. N., 1965, *J. Geophys. Res.* **70**, 3085.

Coppi, B., 1965, *Phys. Fluids* **8**, 2273.

Cowling, T. G., 1933, *Mon. Not. Roy. Astron. Soc.* **94**, 39.

Cowling, T. G., 1965, *Stars and Stellar Systems* **8**, 425.

Drummond, J. E., ed. 1961, *Plasma Physics,* McGraw-Hill, New York.

Dungey, J. W., 1953, *Phil. Mag.* **44**, 724.

Ellis, G. R. A., 1965, *Radio Science* **69D**, 1513.

Fermi, E., 1954, *Astrophys. J.* **119**, 1.

Furth, H. P., Killeen, J. and Rosenbluth, M. N., 1963, *Phys. Fluids* **6**, 459.

Gardner, F. F. and Whiteoak, J. B., 1966, *Ann. Rev. Astron. Astrophys.* **4**, 245.

Ginzburg, V. L. and Ozernoi, L. M., 1966, *Radiofizika,* **9**, 2, 221.

Ginzburg, V. L. and Syrovatskii, S. I., 1964, *The Origin of Cosmic Rays,* Pergamon Press, Oxford.

Ginzburg, V. L. and Syrovatskii, S. I., 1965, *Ann. Rev. Astron. Astrophys.* **3**, 297; 1969, *Ann. Rev. Astron. Astrophys.* **7**.

Ginzburg, V. L. and Zhelezniakov, V. V., 1958, *Astron. Zh.* **35**, 694.

Giovanelli, R. G., 1946, *Nature (London)* **158**, 81.

Giovanelli, R. G., 1948, *Mon. Not. Roy. Astron. Soc.* **198**, 163

Gold, T., 1959, *J. Geophys. Res.* **64**, 1219.

Gold, T. and Hoyle, F., 1960, *Mon. Not. Roy. Astron. Soc.* **120**, 89.

Hayakawa, S., Nishimura, J., Obayashi, H. and Sato, H., 1964, *Progr. Theoret. Phys. Suppl.* No. 30, 86.

Herzenberg, A., 1958, *Phil. Trans. Roy. Soc.* A **250**, 543.

Kiepenheuer, K. O., 1950, *Phys. Rev.* **79**, 738.

McCray, R., 1966, *Science* **154**, 1320.

Melrose, D. B., 1967, *Planet Space Sci.* **15**, 381.

Parker, E. N., 1958, *Phys. Rev.* **109**, 1328.

Parker, E. N., 1963, *Astrophys. J. Suppl. Ser.* **8**, 177.

Petschek, H. E., 1964, *Physics of Solar Flares,* W. N. Hess, ed., AAS-NASA.

Piddington, J. H., 1960, *Geophys. J. Roy. Astron. Soc.* **3**, 314.

Piddington, J. H., 1962, *Nature (London)* **194**, 962.

Piddington, J. H., 1967, *Mon. Not. Roy. Astron. Soc.* **136**, 165.

Piddington, J. H., 1968, *Earth's Particles and Fields*, B. M. McCormac, ed., Reinhold, New York.
Sonnerup, B. U. Ö. and Laird, M. J., 1963, *J. Geophys. Res.* **68**, 131.
Sturrock, P. A. and Coppi, B., 1966, *Astrophys. J.* **143**, 3.
Swann, W. F. G., 1933, *Phys. Rev.* **43**, 217.
Sweet P. A. 1958, *Nuovo Cim. Suppl.* **8**, Ser. 10, 188.
Swift, D. W., 1967, *Planet. Space Sci.* **15**, 835 and 1225.
Zheleznyakov, V. V., 1966, *Zh. ETF,* **51**, 570; *Soviet Phys. JETP,* **24**, 381.
Zheleznyakov, V. V., 1967, *Astronom. Zh.* **44**, 42; *Soviet Astron. AJ* **11**, 33.

CHAPTER 5

The Sun and Solar Activity

The sun is a star of only average luminosity and surface brightness, and one of about 10^{11} in our galaxy. Its interest lies, of course, in its proximity and consequent profound effect on our environment. This proximity also allows it to be studied in vastly greater detail than the next nearest star and so its optical, thermal, radio and x radiations are monitored continuously and its numerous corpuscular radiations are studied whenever and however possible; the latter comprise the solar wind and various more energetic particles. So important are the studies of solar radiations and their geophysical effects, particularly their effects on the earth's ionosphere, that three international scientific unions have formed the Inter-Union Commission on Solar-Terrestrial Physics. The unions concerned are the International Astronomical Union, the International Union of Geodesy and Geophysics and the International Scientific Radio Union. The new commission replaces an earlier one concerned with solar-terrestrial relationships and follows the successful ventures, the International Geophysical Year and the International Years of the Quiet Sun (Stickland, 1968).

The visible sun has a radius of about 7×10^5km (1 $R_\odot$), but its atmosphere extends very much further. The atmosphere might be defined as that part of the solar plasma from which we receive direct optical radiation; on this basis it extends more than 10 $R_\odot$. As we shall see in Section 6.2, however, the solar atmosphere is continuously expanding and may be measured directly in the vicinity of the earth (about 200 $R_\odot$). At the bottom of the atmosphere lies the photosphere and immediately above, the chromosphere, where most optical line emission and absorption occurs, and the highest frequency radio thermal emission takes place. In the lowest 500 km of the chromosphere occurs the lowest temperature of all the solar plasma, about 4700K. There the density of neutral hydrogen reaches a peak of about $10^{16}/cm^3$ and virtually all electrons present come from singly ionized metal atoms.

74

The electron density is only $10^{12}/cm^3$, but this is still adequate to constitute an electrically conducting plasma.

Above and below this relatively cool layer, the temperature rises rapidly, to 10^6K or more in the corona and many times that in the center of the sun. At a level of about 1500 km the picture of a spherically symmetrical atmosphere changes to one of innumerable spicules or jets which project upwards to about 10^4km. The gas temperature inside the spicules is only about 10^4K, but that between them exceeds 3×10^4K. Their identity is preserved by magnetic fields of strength about 10 gauss which are concentrated within the spicules. The spicules themselves and their magnetic fields are related to the convection patterns which are responsible for the transport of most heat from below the photosphere. These convective motions are evidenced by the *granulation*, or small-scale (about 2000 km) mottling, observed when the surface field is weak enough, and by the *supergranulation* or larger-scale pattern associated with stronger fields and with spicules.

At heights of about 12,000 km above the photosphere the spicules have disappeared and we enter the lower corona where the temperature is about half a million degrees. In a distance of about 2% of a solar radius the temperature has risen from about 5000K to about 5×10^5K; beyond that level it is remarkably steady out to several solar radii, but the electron and proton densities fall from about $10^9/cm^3$ to $10/cm^3$ near the earth.

This very brief review of the solar plasma is given as a basis for the discussion of solar electrodynamic phenomena in the following sections. For detailed reviews, the texts of Billings (1966) and Shklovskii (1965) on the corona, Zirin (1966) on general solar physics, and Bray and Loughhead (1967) on the granulation, should be consulted.

The intensive study of *solar activity* over the past two decades has introduced a number of new observational techniques and a great deal of observational data. The *coronagraph*, invented by Lyot, has been greatly improved, and allows observations of the corona outside eclipses. From the complex configurations of the coronal plasma some idea of the forms of frozen-in magnetic fields may be inferred. The work of Babcock (1963) and others had led to a steady improvement of *magnetographs*, so that solar fields of strengths down to one gauss or less may now be measured. Methods of measuring the direction, as well as the strength, of solar fields have also been devised. The development of large and complex *radio telescopes* has revealed the existence of a wide variety of emissions, both steady and transient, at wavelengths from millimeters to decameters. It is possible also to measure the locations and sizes of the sources of these emissions and to identify them with rapidly moving plasma clouds (Wild *et al.*, 1963). *Solar x rays* may now be measured more or less continuously by detectors carried above our absorbing atmosphere in satellites and space probes. Similar techniques are used for the detection of solar ultraviolet emission and both these results have been reviewed by Goldberg (1967). Finally, there is solar *particle astronomy* which has grown to be a major section of solar astronomy.

From these and earlier results stem many problems, most of which involve electrodynamic effects of one sort or another. The basic problem is the origin of the sun itself and of the magnetic field which it must have possessed at that time. Since the sun is part of the solar system and is also a star, these questions are discussed in Chapter 10 (on the solar system) and Chapter 11 (on stars).

Magnetic fields in the photosphere and corona show a complex and ever-changing pattern, which is the subject of Section 5.1. Reviews of solar magnetic fields have been given by Babcock (1963), Severny (1964), several contributors in Lüst (1965) and by Howard (1967). A thorough review of sunspot fields has been given by Bray and Loughhead (1964). It seems convenient to divide these fields into small-scale and large-scale fields, the former extending from the smallest resolvable structure (a few hundred kilometers) to a spot group and its environs (of order 10^5 km); the latter extends to the "general" field of the sun. The problems of the first group involve the physics of granulation, faculae, spots and prominences, while those of the latter center on the general field, its 22 year cycle and its extension into interplanetary space. Kinematical hydromagnetic theory of plasma motions and field development has been developed by Parker (1963) for a number of these problems.

In Section 5.2 the hot, expanding corona is discussed, with emphasis placed on the role of magnetic fields in providing thermal and mass kinetic energy to the corona. A review of the physics of the corona has been given by Newkirk (1967).

Sections 5.3–5.5 are devoted to a discussion of solar activity in its more violent forms. An *active region* develops with its *spot group*, *faculae* and various coronal effects. The development culminates in the most spectacular of all solar phenomena, the *flare event*, after which the general activity subsides. These phenomena have been reviewed by Smith and Smith (1963), contributors in Hess (1964), Tandberg-Hanssen (1967) and Svestka (1967).

The chapter concludes with a brief review of solar radio and particle emissions, given in much more detail by Kundu (1965) and Fichtel and McDonald (1967) respectively.

5.1   Magnetic Fields

The basis of solar electrodynamic phenomena is the complex, varying pattern of magnetic fields. After Hale's discovery of sunspot fields, about 40 years elapsed before the much weaker fields outside the spots were measured (Babcock and Babcock, 1955). These are measured near the surface of the sun, but their configuration in the corona may be inferred from plasma distributions and motions, together with magnetohydrodynamic theory. The fields are known to extend out past the earth's orbit and are measured directly by spacecraft in the vicinity of the earth. Reviews of these results are referred to above, and here we need only repeat those needed in discussing electrodynamic phenomena of interest.

Sunspot Fields

Sunspot fields are the strongest, ranging beyond 3000 gauss and having total spot fluxes of 10^{21}–10^{22} gauss cm^2 in medium to large spots. The appearance of spots is preceded by a strengthening of the weak photospheric field and by enhanced optical line emission in a limited region called a *plage* or *facula*. Somewhere in this plage *pores*, or small regions darker than their surroundings, appear. These contain fields of about 100 gauss and they may amalgamate to form a *spot*.

The three main problems concerning sunspots are, first, how was the magnetic flux created, second, how did it get into the spot, and third, what makes the spot dark. The energy of a spot field is so great that its source must lie well below the photosphere, where adequate hydrodynamic energy is available to build up or amplify a weak field in the manner described in Section 4.1. On the other hand the source cannot lie deep within the sun or the field could not be convected to the surface; all known solar magnetic phenomena originate in a surface layer of thickness only a small fraction of a solar radius. It is believed that these sub-surface fields are convected into place by large-scale hydromagnetic motions. These are driven, in part at least, by magnetic buoyancy (Cowling, 1946); magnetic pressure in a submerged magnetic rope causes a reduction in plasma pressure and density and so the rope tends to float to the surface. This effect may be aided to some extent by magnetic diffusion and, near the surface, by a third effect which may be referred to as a "tearing" of the surface. Arched lines of force are thrust above the initial surface by magnetic stresses and the material elevated with them then slides down the sloping parts of the field lines to uncover new surface material (Parker, 1955). The magnetic buoyancy and tearing effect may be recognized respectively as interchange motion and Rayleigh-Taylor instability, both discussed in Chapter 4.

Spot fields grow and decay in periods of 10–50 days and, as Cowling has shown, such speed implies that the lines of force are convected into place, rather than diffusing through the plasma. Diffusion time is given by equation 2.12 as $t \sim 4\pi\sigma L^2$ and for a spot of dimensions 3000 km and conductivity $\sigma = 3 \times 10^{-8}$ emu this would be about 1000 years. Bray and Loughhead (1964) reduced L to 1500 km, which is the scale of the fine structure of the plasma, and put $\sigma = 10^{-10}$ emu, which gave a time scale of some months. More recently Oster (1968) has found a much lower value of σ in the surface layer of the photosphere (about 10^{-12} emu), which would reduce the time to a few days. However, at slightly lower levels (optical depth 10) the conductivity has risen to about 10^{-9} emu and the field is frozen-in for periods of years. It would seem, therefore, that spot fields must be convected into place, because they are generally directed vertically and are frozen into all but a very thin horizontal sheet of plasma.

The major uncertainty in deciding between convection and diffusion is the distance scale L, which depends to some extent on the field configuration. For example, if a generally parallel set of field lines forms a "rope" which is divided into a number of "strands" (of stronger field), or if it develops small-scale kinks, these features do not determine the value of L for the rope itself. Diffusion may wipe out the small features but it still leaves the rope, whose time scale depends on the value of L for the whole rope. Even after a spot disappears, apparently due to the expansion of its rope of magnetic field lines, the field still exists. Furthermore, as seen below, it may have been stretched out into interplanetary space where it exists for many months, corotating with the sun. Near the earth its strength is typically 10^{-4} gauss and if this value is projected back to the sun according to an inverse-square law, it gives a photospheric field of 4 gauss which is typical of large-scale photospheric fields.

Theories of cooling of sunspots depend on the inhibition of convection by magnetic fields (Biermann, 1941). A cellular convection pattern requires vertical and horizontal motions of plasma in order to convect heat to the photosphere. The field allows the former but if it is strong enough, it inhibits the latter and so blocks the flow of heat and lowers the surface temperature. This idea of spot darkening has been extended by Cowling (1957), who found the field strength needed to inhibit convection of a particular wavelength. The balance of pressures inside and outside the cooled spot has been described in Section 2.2.

Large-Scale Photospheric Fields

Large-scale photospheric fields are observed, having areas of order 10^3 times those of spots and strengths 10^3 times weaker—and hence fluxes of the same order. These fields may be isolated (unipolar magnetic regions) or may appear as related pairs with opposite magnetic polarities (bipolar magnetic regions).

The main problems concerning these fields are two: their relationships to internal fields and to coronal fields, and their relationships to plasma configurations and motions.

As far as surface phenomena are concerned the fields are related to the convection patterns. The basic pattern is shown by the granules, of size about 2000 km, which are distributed rather evenly over the surface of the sun and are not directly related to the large-scale fields. However, a secondary pattern is provided by the supergranulation, of size about 30,000 km, which is found in regions of enhanced line emission. It is believed that the large-scale fields are related to this pattern and that the field strength is greatest between the supergranules, where the chromospheric spicules also originate as jets of hotter gas. For a review of spicules see Beckers (1968).

In the corona the effects of large-scale (and also spot) fields on the plasma is clearly evidenced in photographs or sketches such as Figure 1.1, which was used to illustrate polar plumes and the "general" magnetic field inferred from their configuration. Also illustrated in this composite sketch are the very common coronal features:

(1) Prominences (P) and loops (L) extending up to about one solar radius ($R_\odot$).
(2) Streamers (S) extending beyond 10 $R_\odot$.
(3) Condensations (C) extending up to 0.1 $R_\odot$.

These terms are descriptive, and there is not general agreement about the phenomena referred to. Sometimes the name is changed when a particular structure becomes more active, and in particular prominences appear in many forms and have been given various titles. However, the main distinction here is between quiescent and moving prominences and the latter, together with streamers and condensations, are discussed in Sections 5.3–5.5 as parts of a center of activity.

Numerous models of *quiescent prominences* have been proposed (Tandberg-Hanssen, 1967) but it seems fairly clear that the main factor involved is the support (or even upward acceleration) of plasma by magnetic fields. In the absence of acceleration the problem reduces to the solution of the magnetohydrostatic equation 2.14, first attempted by

Menzel. Perhaps the most successful model is that of Kippenhahn and Schlüter (1957). However, since mass motion may be observed in most prominences, static models only provide an approximation, although probably a reasonable one in the case of quiescent prominences.

The relationship between the surface field and the fields below and above the solar surface may commonly have the form illustrated in Figure 5.1. A twisted magnetic rope has been generated in the interior

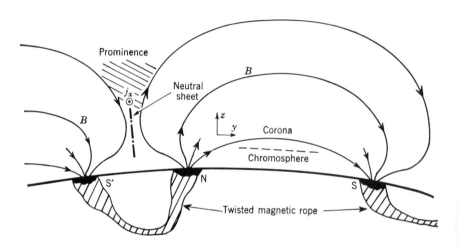

Fig. 5.1. A vertical section through the sun and three spots, to show a possible magnetic field configuration. A sub-surface magnetic "rope" has emerged between the spots N and S which form a bipolar pair, and to the left of spot S'. N and S are about 10^5 km apart and the closer pair NS' provide a magnetic neutral sheet (dot-dash line) and pinched current j_x out from the paper. A prominence may be supported by the field in the region shown hatched.

of the sun in the manner described in Section 4.1. This rope, which originally lay more or less parallel to the surface of the sun, has burst through the surface in the region between the spots N and S and also to the left of spot S'. The forces responsible for this breakthrough are magnetic buoyancy and instability caused by excessive twisting. The field above the surface expands into the corona to levels of 10^5 km as shown, or more. The configuration, and the subsequent development of *eruptive prominences*, streamers and condensations are features of an active region and as such are discussed in Sections 5.3–5.5.

General Magnetic Field

The existence of a general magnetic field of the sun may be inferred from measurements of a weak field (about 1 gauss) around both poles and down to latitudes ±55°. The polar plumes of Figure 1.1 suggest a dipole-type field and Babcock (1961) has proposed a model based on earlier discussions of Cowling and of Allen (1960), to explain such a field which reverses every 22 years. The model also explains the observed latitude drift of sunspots and the reversal of the polarity of the leading (westward) spot of a pair of spots, each 11-year period.

The essential elements of Babcock's theory are as follows: First, the lines of force of a poloidal field are drawn out by differential rotation to form a much stronger toroidal field as shown in Figure 4.2. If the north pole is a north magnetic pole, then the leading spot in the northern hemisphere is also north magnetic and that in the southern hemisphere is south magnetic, as shown in Figure 5.2. As magnetic buoyancy lifts the toroidal flux ropes, coriolis forces acting on the plasma rising in convection cells cause following spots to appear in higher latitudes than leading spots. The magnetic flux loops then appear as loop R. A pair of loops S and T from two spot pairs as shown are separated by a magnetic neutral sheet X near the equatorial plane. A line of force

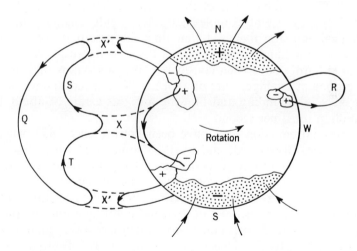

Fig. 5.2. Babcock's model of the solar magnetic oscillator which provides a general (poloidal) field reversing every 11 years. Bipolar magnetic areas, one in the northern and one in the southern hemispheres, provide loops (such as that shown separately as R) which lie in meridian planes. Two loops S and T join one another across the neutral sheet X. They also neutralize the general field, across neutral sheets X′, X′, and then reverse it. A large magnetic loop Q is released into the corona.

of the general field Q is also separated from the two spot loops by magnetic neutral sheets X'. Field annihilation and reconnection in these three neutral sheets remove the dashed portions of the field lines and replace them with the parts of the field lines which connect the ends of the dashed lines. The inner loops then contract back to the sun and the large outer loop is released into the corona. In this way the original poloidal field is wiped out, and then continuing field reconnection in the region X generates a new poloidal field in the *reverse sense*.

Babcock's theory requires considerable extension to provide complex subphotospheric motions required to account for all features of the magnetic oscillator. Also, it may be too simple to account for all more recent observations of solar fields. However, it does illustrate the various effects involved and may well be capable of extension and improvement. Attempts in this direction have been made by Chvojkova (1965), Vostŕy (1967) and Leighton (1968).

5.2 The Hot Corona

In 1933 Rosseland observed that the solar corona "has stimulated speculation to the breaking point," until some authors had even suggested that "the familiar laws of nature are set at nought in the corona." The characteristics of the corona which evaded explanation included its great extent, complex and often rapidly changing structure, and optical emission lines. At that time the possibility of a coronal temperature one hundred times greater than the photospheric temperature could hardly have been considered. We now know that not only is the corona at a temperature of about 10^6K, but that the whole solar atmosphere is expanding and blowing away at a rate of about 10^{12}g (one million tons) per second.

The coronal emission lines have been known since 1869, when the green line was discovered, and by 1939 some twenty-four lines were known. Throughout those years fruitless attempts were made to identify these lines, but the first real progress was made by Grotrian (1939), who ascribed two lines to FeX and FeXI. In 1942 Edlén identified almost all known lines as transitions in even more strongly ionized atoms, a result which provided fairly conclusive evidence of a very high temperature. Further optical evidence of a temperature of order 10^6K is provided by the width of the coronal emission lines and the obliteration of the Fraunhofer lines (originating in the chromosphere) in the light scattered by coronal electrons.

Independent, and equally conclusive, evidence of a hot corona is provided by the steady thermal radio emission. At centimeter wavelengths the corona is optically thin and so by Kirchoff's Law its emissivity is small and most of the radiation comes from the cooler chromosphere. As the wavelength increases, the emission source moves up into the corona and the brightness temperature increases towards 10^6K, which is the black body coronal emission. The theory of such varying radiation was worked out by Martyn (1946) and Ginzburg (1946). Later, the full range of radio data, together with optical data relating density and temperature, was used to give model atmospheres averaged over the disk (Piddington, 1954). Unfortunately, the conditions vary greatly from spicules to interspicule regions and only a weighted average is obtained. However, the method gives a simple and powerful test of any model, however complex, and should be used on all new models. While there was still some doubt concerning the temperature of the corona, Alfvén pointed out that the electron density distribution determined by Allen and van der Hulst also indicated a temperature of about 10^6K. This value is obtained on the assumption of an equal density of protons and electrons and the condition of hydrostatic equilibrium; all observational evidence was thus in general agreement.

The precise mechanism of coronal heating is not known, but it is generally accepted that thermal energy derives from mass kinetic energy of waves which originate below the photosphere and move into the corona. A second source of energy, which may be even more important in active regions, is conversion of magnetic to gas kinetic energy in magnetic neutral sheets, as described in Section 4.2.

It appears to be the spicules which provide the flow of energy into the corona, over the surface of the quiet sun. These provide mass motions into the·corona up to their peak levels of about 10^4 km where they have two effects. The spicule atoms collide with the atoms of the corona and heat the latter; the spicule mass motions create waves which propagate into the corona where they dissipate energy. These waves may be *acoustic, gravity,* or *hydromagnetic waves,* and here we are mainly concerned with the possible significance of hydromagnetic waves, first discussed by Alfvén (1947). In the early discussions of energy dissipation of weak waves in the corona, the effective electrical conductivity was greatly underestimated; when this was corrected the heating rate was found to be quite inadequate. In the chromosphere, energy is dissipated much more rapidly because of collisions between ions and neutral atoms which may be very severe. However, the upward revision of the collision cross section, and also of the frequency of the waves carrying the greater part of the energy of convective motion, indicated that heating by this mechanism is not likely to be the major contribution (Osterbrock, 1961).

The important effect may be the transmission of the longitudinal plasma (sound) wave through the chromosphere (where the transverse Alfvén modes are absorbed). The wave becomes increasingly magneto-hydrodynamic as the ratio of magnetic stress to pressure gradient increases, and it also becomes more violent as the density decreases. Finally it becomes a shock wave, which dissipates rapidly and heats the upper chromosphere. Osterbrock found that at great heights, where the magnetic field dominates, the shocks become weaker again and are refracted back down towards the photosphere. An important effect at high levels may be collisions between shocks and generation of the two transverse magnetohydrodynamic modes. These travel along the magnetic field lines and may transport energy freely into the corona. Pikelner and Livshitz (1964) discussed the variety of magneto-hydrodynamic waves originating in different granulae. In the middle and upper chromosphere the lengths of all three wave types is comparable with the sizes of the flat portions of their fronts so that an approximate treatment using geometrical optics is not valid. Lüst and Scholar (1966) also found a radial steepening of wave fronts and a strong tendency for waves to be channeled along the magnetic field lines. Uchida (1968) explained blast waves, observed to propagate across the solar disk during some flares, as the intersecting line between an expanding coronal wavefront and the chromosphere.

5.3   Morphology of Active Regions

Solar active regions or centers of activity, as the names imply, are the sites of a variety of phenomena, observed in the photosphere, chromosphere and corona. As the Babcocks and others have shown, the growth and decay of active regions is very much a history of magnetic configurations, although these must be related to the optical features, notably the supergranulation (Leighton, 1964; Bumba and Howard, 1965). Histories of active regions or centers of activity have been given by Kiepenheuer (in Xanthakis, 1967 and earlier), de Jager (1965 and earlier), Bumba and Howard (1965) and Pikelner (1965).

An active region first appears as a developing magnetic field, preferably within or close to an old expanding magnetic region whose field has fallen to about 1 gauss or less. The region may have an extent up to a few times 10^5 km over the photosphere, and upwards far into the corona. If the magnetic field strength exceeds about 10 gauss, then this "magnetic plage" provides enhanced optical emission, particularly line emission such as $H\alpha$ and the CaII, K line. The latter propagates around the borders of adjacent supergranules and develops into a plage or facula. The optical plage also closely defines a radio plage or region of enhanced, slowly varying emission at wavelengths below about 30 cm. This emission is thermal (Piddington and Minnett, 1951), originating in a condensation or region of higher density (for a particular value of temperature) or higher temperature (for a particular density). More recently, ultraviolet plages and x-ray plages have been discovered, also lying above the optical plage and also probably of thermal origin (Elwert, in Hess, 1964).

The reason for the higher chromospheric temperatures in a plage is modification of the convection pattern by the magnetic field. At some depth below the photosphere energy is transmitted upwards mainly by convection, and the flow is limited by a viscous effect due to turbulence. Gas motions in the convecting cells are to some degree random, and ascending and descending currents interfere with and retard one another near their borders. A magnetic field of 10–100 gauss is capable of suppressing turbulence and so speeding up the convective motion and heating the photosphere. Maximum heating is obtained when the field is just strong enough to suppress turbulence, and further increase then has the opposite effect because it begins to suppress the convection itself. As seen above, when the field strength rises above about 100 gauss pores begin to form and when it reaches about 500 gauss these coalesce to form a spot. The field has other effects of course, discussed in the preceding section in connection with coronal heating.

Sometimes the plages which first define the heating of an active region also define the maximum activity of such regions. However, for fields of a few hundred gauss or more, a spot, or a group of spots, develops and the extent of the activity is greatly increased. At this stage the magnetic configuration might have the form shown in Figure 5.1. The sub-surface field has the form of a twisted rope and the external field adjusts itself so that Lorentz forces are small (Section 2.2). We know that part of the surface magnetic flux was there before the birth of the active region. Hence the most important effect in spot formation is that a thin part of the sub-surface rope should rise towards the surface. The effect is then to gather field lines, some of which have already passed through the surface, into a confined area which becomes a spot. Other field lines have not previously passed through the surface so that a second important effect is the introduction of additional flux into the area. This cannot be achieved by magnetic diffusion but rather by magnetic buoyancy aided by the Rayleigh-Taylor instability. The magnetic pressure in a magnetic rope causes a reduction in plasma pressure and density, the rope bends and projects through the surface with plasma flowing along the field lines down to sub-surface levels.

With the appearance of spots the activity takes new forms. In the chromosphere long threads may appear as bright objects against the disk in Hα filtergrams. These may extend from 2×10^4–10^5 km between the penumbrae of spots of opposite polarity. Sometimes these threads appear in bundles and sometimes they undulate; sometimes they have loop formation and seem to be identified with magnetic flux tubes extending from one spot to another. Such a tube would correspond approximately to the lowest of the three lines of force connecting spots N and S in Figure 5.1. Seen on the limb, threads and loops differ in that the latter may extend into the corona as shown by L in Figure 1.1. If they are large and well enough defined they may be classified as prominences.

A common feature of the solar atmosphere, and one having a wide variety of forms, is the prominence, an example of which is shown as P in Figure 1.1. *Active prominences* are generally associated with flares and may include effects such as plasma streaming up or down paths which are apparently defined by a magnetic field. In its extreme form, an eruptive prominence expands and ascends with uniform velocity of several hundred kilometers per second. A possible model of an eruptive prominence is described below.

Quiescent prominences take the form of enormous sheets of relatively cool plasma standing vertically in the solar atmosphere. These sheets may be 2×10^5 km long and 5×10^4 km high, with a thickness of only about 5000 km. Seen on the disk they are long dark filaments. Smaller dark filaments also occur in active regions, occupying the dividing line of opposite polarities in a bipolar magnetic region. Hence they are orientated perpendicular to the paper in Figure 5.1. One possible model of quiescent prominences comprises plasma which is supported in the V shape between the magnetic fields of two spots of opposite polarity which are not magnetically linked above the photosphere. Spots N and S′ form such a pair and the prominence or filament is shown in section by the hatched area.

If an active region develops sufficiently large, and preferably complex, spot groups, then activity may culminate in a flare event. This event, or a series of such events, may occur during the spot phase of an active region and is described in the following section. Most theories of flare events depend on the development of a magnetic neutral sheet between oppositely directed fields. Our description of Figure 5.1 is completed by reference to such a sheet which is shown in section by the dot-dash line between spots N and S′. The neutral sheet is also a current sheet, denoted by j_x and flowing out from the paper. Effects which may occur in such a sheet have been described in Section 4.2.

5.4 The Flare Event

An *optical flare* is the temporary enhancement of line emission from parts of a plage area. It is an extremely interesting, but not very impressive, phenomenon. The power radiated in the optical spectrum only increases by about 1% above normal photospheric emission. On the other hand the associated effects, which are referred to as the flare event, are extremely impressive because of the exotic forms of energy represented: x rays, radio bursts, cosmic rays, ejections of plasma into the corona and an interplanetary blast of plasma. However, in spite of this impressive array of flare phenomena, perhaps 90% or more of all observations and interpretations have been done in the Hα line. The result has been a tendency to overemphasize the significance of the optical event and the development of models in which the energy release or "explosion" occurs in the chromosphere. As we shall see, the evidence now points to the flare event occurring far above the optical flare region, the optical emission being the result of bombardment of the denser chromosphere by fast particles moving down the field lines. This model is similar to the generally accepted model of a *terrestrial aurora*.

Even if they are only secondary effects, optical flares are extremely important sources of information, and we reproduce a flare photograph in Figure 5.3. This will be referred to during the discussion of flare characteristics, but it should be borne in mind that the source of emission is deep in the chromosphere where the gas temperature is relatively low (about $10^4 K$ or 1 eV). This may be compared with the "temperatures" characterizing effects such as x-ray and cosmic-ray emissions, which range from $10^8 K$ to $10^{13} K$ or more.

In any discussion of flare-event models the following seem to be the features which must be taken into account:

(1) The optical flare occurs in a contorted region among visible spots of opposite polarity. A spot is visible in Figure 5.3 (partly covered) at the top left of the contorted region; smaller spots are obscured by the projections at the lower right of the flare. This flare, like flares in general, grew out of a plage, expanding and twisting, in a period of about half an hour. A common flare pattern has the form of two twisted "ribbons" lying more or less parallel, as shown in Figure 5.4a and perpendicular to the line joining the spots (shown as circles). Sometimes the two ribbons merge, but as the flare develops they move apart, as shown in Figure 5.4b, until they touch the spots. There is evidence of two flare ribbons in Figure 5.3, one partly covering a spot. Sometimes a quiescent prominence, or dark filament, lies between the flare

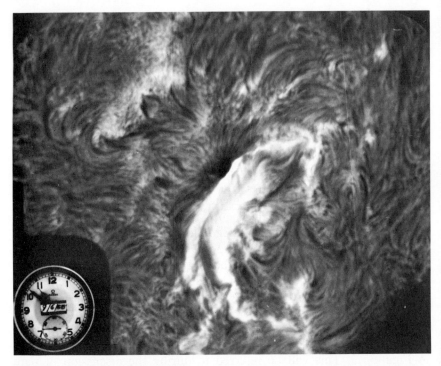

Fig. 5.3. A photograph of a solar flare taken at C.S.I.R.O. Optical Observatory, Culgoora, N.S.W., Australia, on June 9, 1968. A half-obscured sunspot lies under the top left-hand side of the flare.

ribbons, as shown in Figure 5.4c (Kiepenheuer, in Hess, 1964). A review of optical observations of flares has been given by Svestka (1966).

(2) Hydrodynamic blast disturbances sometimes emanate from the site of a visible flare and move across the disk with velocities in the range 1000–2000 km/sec (Athay and Moreton, 1961). They are presumably responsible for the activation of prominences at distances up to 10^6 km (Moreton, 1964; Wild *et al.*, 1968).

(3) A blast of similar velocity moves away from the sun and causes geophysical disturbances 1–2 days later. Various ejections of plasma into the corona may be observed in coronal pictures similar to that of Figure 1.1.

(4) Somewhere within the active region there are accelerated relativistic protons and electrons. The former are regularly observed near the earth where their energy density may reach 100 times that of galactic cosmic rays. The latter usually remain trapped and their synchrotron emission is responsible for some of the radio bursts.

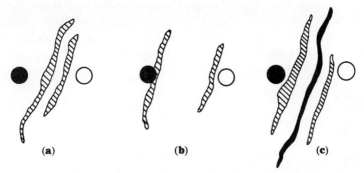

Fig. 5.4. A flare-sunspot configuration sometimes observed. (a) At an early stage, two flare ribbons are close together or merged. They lie between the two spots (full and open circle) roughly perpendicular to the line joining the spots. (b) Later the two flare ribbons have drawn apart until they touch or overlap the spots. (c) Sometimes a quiescent prominence (seen on the disk as a dark filament) lies between the flare ribbons.

(5) There is also copious emission of ultraviolet and x-ray quanta, some by bremsstrahlung and some by synchrotron emission of electrons in strong magnetic fields.

Detailed case histories of particular flares and of averages of many flares have been given by Bruzek (in Hess, 1964 and in Xanthakis, 1967). A quantity of basic importance in any flare theory is total energy expended on all of the flare phenomena. Bruzek has collected order of magnitude estimates of the energy expended in the different radiations from a very large (class 3 +) flare, and these are given here in Table 5.1. The total number of particles involved in a particular effect is N, and their mass is given in the last column.

Bruzek points out that the total number of particles involved in the optical emission is 10^{41}, or about $10^{22}/cm^2$ over the surface of the disk. The total population of the corona is only $10^{19}/cm^2$, so that such a mass could not be condensed from the corona. He concludes that "this is a strong argument for the assumption that the bottom (and the origin) of the flare lie deep in the chromosphere." It would seem, on the contrary, that the data of Table 5.1 provide evidence that the region of energy release cannot lie deep in the chromosphere. Otherwise, of the total mass affected (2×10^{17} g), more than 1% (2×10^{15} g) would reach interplanetary space, carrying with it a major proportion of the total kinetic energy released. It seems more likely that the flare event occurs at a higher level where a smaller mass would receive the energy, thus accounting for the energy-to-mass ratio of the interplanetary blast which is about 100 times that of the optical flare region.

Table 5.1.   Flare Energy, Particle Number and Mass

Flare effect	Energy, erg	Particle number, N	Mass, g
Hα	10^{31}		
Total line emission	5×10^{31}		
Continuum emission	8×10^{31}		
Total optical emission	10^{32}		
Optical flare region		10^{41}	2×10^{17}
Soft x-rays (1–20 Å)	2×10^{30}		
Energetic x-rays (50 keV electrons)	5×10^{31}	10^{39}	
Type IV burst (3 MeV electrons)	5×10^{27}	10^{33}	
Type III burst ($\gtrsim 100$ keV electrons)	10^{28}	10^{35}	
Visible ejection ($v \sim 3 \times 10^{7}$ cm/sec)	10^{31}	10^{40}	2×10^{16}
Energetic protons (E > 10 MeV)	2×10^{31}	10^{35}	
Cosmic rays (1–30 BeV)	3×10^{31}		
Interplanetary blast	2×10^{32}	10^{39}	2×10^{15}
Moreton blast	$> 10^{30}$		

It is generally accepted that the energy source is magnetic and if we consider a more common flare, of energy say 10^{32} erg, then we need a field of 100 gauss occupying a cube of side 6×10^{4} km. Mechanisms for converting magnetic to kinetic energy have been discussed in Section 4.2, and if one of these is assumed to operate, then the effects on the field of Figure 5.1 may be determined. If the spots N and S' have fields of say 2000 gauss, then the strength and extent of the field near the neutral sheet is adequate to meet the energy requirement. Sometimes within complex spot groups the spots are closer, even merging, in which case the observed explosion is more violent (a "proton flare").

Suppose that magnetic field annihilation occurs within a limited region of the neutral sheet, say in a narrow strip more or less perpendicular to the plane of the paper. Above and below this strip the field lines connect across the neutral sheet and because these lines are very severely bent they tend to contract and straighten themselves as shown in Figure 5.5. Field lines below the annihilation strip contract downwards to join the two spots (lines B_1), while field lines above the strip contract upwards and link two magnetic regions far to the left and right of the spots (lines B_2). This process of contraction completes the annihilation of a pair of sheets of magnetic field on either side of the neutral sheet. Two fresh sheets are then pressed together and the process is repeated until the magnetic configuration has changed to that where all field lines belong to the systems B_1 or B_2.

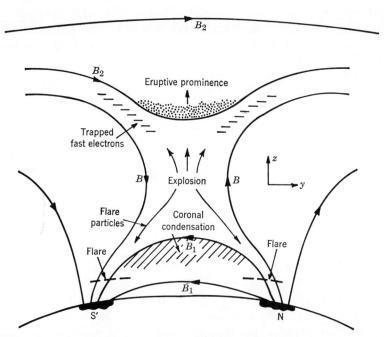

Fig. 5.5. The magnetic field configuration which has developed near the two spots NS′ (Figure 5.1) after a flare event. Field lines have connected across the neutral sheet as in Figure 4.3. The bottom sections have contracted downwards to form the loops B_1 and a coronal condensation and two optical flare ribbons (lying perpendicular to the paper). The top sections have drawn upwards to provide an eruptive prominence and interplanetary blast wave.

Reconnected field lines below the annihilation region contract downwards towards the chromosphere carrying with them trapped plasma, some of which has been heated in the annihilation region. This plasma is further compressed and heated and is then retained in a dome-like structure by the field lines B_1. This is the coronal condensation which is an important feature of the flare event (de Jager, 1965), and whose hot plasma is a source of thermal radio and x-ray emission.

The faster particles in the downcoming plasma are subjected to first-order Fermi acceleration in the contracting magnetic "bottle." The acceleration tends to be along the field lines and the particles eventually bombard the cooler chromosphere to provide the optical flare. Motion is predominantly downward and the emission is from a narrow strip lying directly underneath the neutral sheet. However, as the linking field lines B_1 multiply, they project above the cooler atmosphere and the downcoming stream of particles is redirected to right and left as shown, to provide a pair of flare ribbons of the form illustrated in Figure 5.4a. This is very similar to the terrestrial aurora whose particles, originating in the distant magnetosphere, are divided and directed to the two hemispheres. As the field lines B_1 continue to accumulate and bulge upwards, the two flare ribbons are separated more and more as in Figure 5.4b.

A quite different set of flare-associated phenomena occurs in the region above the strip where the field lines are reconnected. This section of the field contracts upwards and so gives rise to plasma motions away from the sun. In this way some eruptive prominences and similar effects may be explained as well as the dissipation of some quiescent prominences in the vicinity of flares. Some energetic protons and electrons, originating in the region of the neutral sheet, may also escape outwards to account for solar particle events.

The flare-event model may take two rather distinct forms depending on the level at which magnetic fields reconnect. If this level is near the top of the neutral sheet shown in Figure 5.1, then almost all of the energy provided by the subsequent contraction of the field lines is associated with downward motions of plasma and fast particles. In this case the optical flare and the coronal condensation should be relatively powerful, and x-ray and thermal radio emission should be strong. On the other hand, if the reconnection level is low (as would be expected if spots N and S' are close together, or better still if their fields are merged in one spot), then most of the magnetic energy will be fed to upward moving plasma and particles. In this case the outward motion may be strong enough to initiate an interplanetary blast wave, and the field lines B_2 may be drawn out radially to form a semi-permanent feature of interplanetary space (Piddington, 1958). In this case the emission of fast particles is more likely and, for merged spots, the result may be a proton flare.

According to the field annihilation mechanism of Section 4.2, plasma is accelerated to a very high velocity in the direction of the original current in the neutral sheet. This direction is out from the paper in Figure 5.1, and so the ejected plasma moves more or less parallel to the surface of the sun. This ejection may account for the observed horizontally travelling shock waves and for the triggering of sympathetic flares in nearby neutral sheets.

5.5 The Post-Flare Period

A comparison of the magnetic field structures of Figures 5.1 and 5.5 reveals some major changes which, in turn, may be related to the plasma features shown schematically in Figure 1.1.

First we have the coronal condensation (C of Figure 1.1), which is a volume of hot plasma compressed against the chromosphere by tension in the magnetic field lines B_1 (Figure 5.5). Numerous models of these condensations have been constructed to account for the excess thermal emission in the radio spectrum, in the x-ray spectrum or as coronal optical line emission. A typical condensation might have a diameter of about 10^5 km and be several times as bright in coronal yellow line (Ca XV) emission as the normal corona. This brightness requires a high central temperature, perhaps as much as 5×10^6K, and plasma densities in excess of the surroundings. Such a model makes two demands on the magnetic field: First, magnetic stresses must be able to contain this bubble of gas whose pressure is several times that of its surroundings. Second, if the condensation is to persist for any length of time it must be thermally insulated. The magnetic field is capable of reducing transverse thermal conductivity and so perhaps of keeping the bubble hot; this is the reverse of the requirement of keeping quiescent prominences cool in a hotter corona. The coronal condensation is related to the plage phenomenon (optical and radio), which is confined to lower levels where the temperature ranges from about 5×10^5K down to slightly above normal chromospheric values. The plage also requires a source of heating and compression, and it also emits thermally. As we have seen above, its source of heating in the pre-flare period is probably by magnetic control of the thermal convection pattern. However, in the post-flare period some heating may occur as a result of field line reconnection, in which case the plage would really be the base of a condensation.

A coronal structure often developed during a flare event is the loop, also shown (L) in Figure 1.1. The yellow-line regions often take the form of this basic geometric form (Billings, 1966). Zirin (1966) has shown how loops feature in the complex interplay between flares and coronal activity. He observed limb events in which strong line emission originated in a coronal plasma cloud just prior to the occurrence of a flare. The loop structure, like the others of Figure 1.1, is strongly suggestive of magnetic control and may perhaps be explained in terms of field lines B_2 of Figure 5.5. Prior to their connection through the neutral sheet these lines connected with spots N and S', but on release they moved outwards to form a large-scale loop.

The third coronal structure illustrated in Figure 1.1 is the streamer (S), which extends far beyond the loops. Observations of the scattering of radio waves from the Crab Nebula as it moves behind the solar corona indicate that the plasma is drawn out into radial filaments for some solar radii. Again this structure is strongly suggestive of magnetic control and the streamers are thought to be defined by magnetic field lines whose outer portions were carried away during the flare explosion, or were subsequently carried away by a strong solar wind originating in the hotter corona. The lines are carried far into interplanetary space and, as seen in the following chapter, become a semi-permanent feature of interplanetary space.

Finally, we should consider the post-flare photospheric magnetic fields. The flare was caused in the first place by the formation of bundles of field lines which then created the spots N and S'. The forces responsible were located below the photosphere and were much greater than any forces available above the photosphere, including those in the flare event. There is no reason, therefore, why the flare should signal the cessation of the sub-surface motions and the growth of the spots. A new neutral sheet might tend to develop and the chromospheric field pattern might revert to the pre-flare form, as sometimes observed (Howard and Babcock, 1960). In this case another flare of the same pattern might result, thus accounting for homologous flares which are not infrequent (Ellison, 1963). We are left with the disposal of the loops of magnetic field B_1 of Figure 5.5; these lie partly above and partly below the photosphere and if allowed to accumulate would eventually lead to excessive photospheric field strength. These loops will tend to destroy themselves by contracting until the field lines vanish in points; they are also likely to be pushed upwards until they lie entirely within the solar atmosphere, in which case their energy must be released as plasma motions.

5.6   Emission of Radio Waves and Energetic Particles

The energy quantum of a solar radio wave is less than 10^{-3} eV while the energy of some ejected particles exceeds 10^{10} eV. In spite of this disparity the two emissions are closely related and both may be accounted for by the magnetic neutral sheet mechanism illustrated in Figures 5.1 and 5.5.

Radio Waves

As usual, the radio emission has negligible dynamical significance, but because of the sensitivity and other qualities of the radio telescopes, it provides a great deal of information about the motions of plasma and groups of particles and so may be used to check flare models. Reviews of the various types of radio emission have been given by

Wild *et al.* (1963) and Kundu (1965), and of the theory by Takakura (1967).

Solar thermal emission is recognized by its steady flux and spectrum (Figure 4.6) and has proved most useful in checking models of the distributions of temperature and density in the solar atmosphere. There is also enhanced thermal emission from an active region (radio plage), and within a few minutes of the onset of a flare microwave ($\gtrsim$ 20 cm or $\lesssim$ 1500 MHz) emission often increases measurably. This shows that the flare is not an unheralded "explosion," but is preceded for some minutes by an increasing chromospheric temperature. At the time of the flare there is a further rapid enhancement called a microwave burst. While this is thermal in the sense that it is due to the random motions of fast particles, it is probably not bremsstrahlung but rather cyclotron radiation by electrons of energy about 10^5 eV in a field of about 10^3 gauss. Coincident with these microwave bursts are x-ray bursts which may also be accounted for by the same electrons, this time as bremsstrahlung from lower levels. A relatively small number of electrons ($\sim 10^{33}$) is able to account for the microwave burst and also the increase in emission in the short-wave tail of the x-ray spectrum ($\gtrsim$ 20Å or $\gtrsim$ 600 eV) by several orders of magnitude.

Some flares have a distinctive flash phase and perhaps half of these are accompanied by the simultaneous emission of radio type III bursts. These have a distinct spectral peak (Figure 4.6, curve b) which is sometimes accompanied by the second harmonic, and drifts rapidly from the highest frequency of about 300 MHz to about 30 MHz. These fast-drift, meter-wave bursts are interpreted in terms of plasma oscillations excited by the Cerenkov mechanism, as clouds of electrons move outwards from the site of the flare event at speeds of 0.2–0.7 c. As we have seen, the plasma waves themselves cannot escape from the sun but they may generate radio waves at the first two harmonics of the plasma frequency by scattering on coronal inhomogenieties. Estimates of the number of streaming electrons required for a typical group of ten type III bursts is about 10^{36} electrons in the energy range 10–200 keV.

Some of the outward travelling electrons escape from the sun, but others are trapped in magnetic loops where they provide type V bursts, perhaps by the Doppler-shifted cyclotron emission mechanism.

Most flares, especially small ones, exhibit only the flash-phase phenomena described above. In some cases a second phase is introduced by a radio "outburst" or type II burst beginning several minutes after type III bursts have subsided. Type II bursts have a slow frequency drift ($\gtrsim$ 1 MHz/sec), their duration is about 10 min and their brightness temperatures are about 10^{11}K. They may show complex spectral forms with harmonics, and sometimes the dynamic spectra (plot of observed flux density in frequency–time coordinates) show secondary bursts similar to type III and issuing from the main type II disturbance. All of these effects are consistent with the "plasma hypothesis" but instead of a fast electron cloud (~ 0.5 c) we must now invoke a slower hydromagnetic shock ($\sim 10^3$ km/sec) which moves outwards and excites successively lower frequencies. The original eruption is now of sufficient energy to move the ions as well as the electrons. The shock propagates normally to the magnetic field and sporadically ejects fast electrons along the field. Thus the model which emerges from radio burst analysis is consistent with that described in Section 4.2 to explain magnetic field annihilation.

Some large type II bursts are followed by continuum radiation which may cover the whole observable radio spectrum and last for hours or even days. There may be two distinct components of this type IV emission. The first is emitted from extensive regions ($\sim 5 \times 10^5$ km) in the corona which have transverse speeds up to 3000 km/sec, and is thought to be synchrotron radiation from electrons of energy about 3 MeV in fields of about 1 gauss, some 4×10^{32} electrons being required. The relatively slow speeds of the electron clouds must be explained as a hydromagnetic effect, involving movements of magnetic field lines and protons as well as electrons. Other type IV bursts originate in smaller regions ($\sim 10^5$ km) fixed on the solar disk and emitted as circularly polarized ordinary mode. This type is believed due to another group of fast electrons trapped in magnetic loops and exciting Cerenkov plasma waves which, in turn, couple with the O mode of the electromagnetic field. The best known indicator of a terrestrial geomagnetic storm is the occurrence on the sun some 30–40 hours previously of a combined type II-IV event.

The type IV spectrum extends more or less continuously into the microwave region. This emission, like the more impulsive microwave bursts, is attributed to cyclotron radiation from electrons in the 100 keV range in magnetic fields of about 1000 gauss. Presumably these electrons are trapped in the low-lying magnetic loops. The significance of this emission is its remarkably close association with the emission of solar protons capable of delayed detection at the earth.

Energetic Particles

Energetic particles, mainly protons, of solar origin have been observed near the earth since 1942 and the data reviewed by Fichtel and

McDonald (1967). A striking feature of solar cosmic-ray events is the great variety observed, ranging from rare events following great flares, when protons of energy more than 20 MeV have flux above 10^3 particles/cm^2/ster/sec for more than a day, and the energy incident equals that of the primary cosmic radiation for a year. At the other extreme are events 10^3 times smaller with no observable associated solar flare. The impulsive emission of electrons of energy greater than 40 keV has also been observed by Van Allen and Krimigis (1965) with detectors on the Mars-bound spacecraft Mariner 4. These events have abrupt onsets, increase to maxima in a few hours and decay over periods of about a day. They are associated with radio and x-ray bursts and other flare-event phenomena.

These various solar particle events were first recorded on sea-level ion chambers, then neutron monitors and later by their ionospheric effects at high latitudes (Chapter 8) which provided greater sensitivity. A great advance was made with balloon observations and then with rockets, satellites and space probes which since 1961 have played a dominant role. As sensitivity increased to reveal protons of energy less than 0.5 MeV, the frequency of events increased. The long-term anisotropies which are observed provide strong evidence of an intermediate storage mechanism at the sun.

Although many weak solar particle events have no associated flare they may well be part of a flare event. All major particle events are associated with flares and, as was seen in Section 5.4, the optical flare is only a single and often minor part of the flare event and one which might well be absent in particular magnetic and plasma configurations. Hence we will consider particle events in general to be another feature of flare events and to provide more evidence of the nature of the mechanism involved. The character of particle events differs remarkably from one event to another. Thus in two events the integral fluxes of energies above 10 MeV have been nearly the same, but above 100 MeV they differed by a factor of 200. In some events the maximum of low-energy intensity occurred 4 hours after the flare, in others 40 hours after. Differences in the abundances of energetic protons, helium nuclei and electrons show striking variations. Proton-to-helium ratios from 1 to 50 have been observed, and changes by factors as large as 5 occur during a single event. The radio observations and theory show that electrons are also accelerated into the MeV range above a flare. Their relative scarcity near the earth is due, at least in part, to trapping near the sun and to synchrotron loss. In fact, it has been shown that the synchrotron radiation from at least one flare event can be explained in terms of electrons whose initial rigidity spectrum is similar to that of the protons and whose number is comparable to that of the protons.

These various characteristics of energetic particles from the sun should provide useful evidence concerning the mechanism of acceleration. Unfortunately it is clear that some of the characteristics are propagation effects acquired after acceleration, and this makes interpretation more difficult. Nevertheless, a reasonably consistent picture now appears and this may be related to the field annihilation model of Section 4.2 and the more general flare-event model of Section 5.4, which is illustrated in Figures 5.1 and 5.5.

We identify the flash phase of a flare event with magnetic field annihilation over a quite limited region of a neutral sheet. The principal effect is electron acceleration within 1 sec and their ejection along the field lines. Outward moving electrons cause type III bursts and inward moving electrons excite microwaves and lower, where the plasma density is about $10^{12}/cm^3$, they cause x-ray bremsstrahlung. The energy involved is relatively small ($\sim 10^{27}$ erg) and the particles are accelerated only to a few times 10^5 eV. However, the short time constant seems to require acceleration by electric fields and to favor electrons rather than protons. This conclusion is consistent with the fact that proton emission does not accompany flare events which have no second or main phase.

The main phase of a flare event involves the development of a shock hydromagnetic wave moving across the field lines, energy being derived from field annihilation. As well as more type III bursts, type II bursts now appear and within a period of about 5 min type IV bursts occur. The latter indicate the presence of electrons in the MeV range and correlate closely with the release of solar protons with similar and sometimes higher energies. Sakurai (1967) and others have shown that the radiating electrons and the solar protons are accelerated in the same region, perhaps by the same process. The ample time between the flash phase and the second phase allows one to accept the Fermi or betatron processes as responsible for the acceleration.

Summarizing the radio and particle data and their theoretical interpretation, we find a total of three heating and accelerating mechanisms. First is the gradual rise of microwave emission due to plasma heating. Second is the flash phase, when packets of electrons of energy about 100 keV are accelerated in an electric field as "runaway" electrons, some ejected outwards and some inwards to cause different emissions. Third is the main phase when hydromagnetic effects predominate, to accelerate particles further to energies exceeding 1 MeV.

References

Alfvén, H., 1947, *Mon. Not. Roy. Astron. Soc.* **197**, 211.

Allen, C. W., 1960, *Observatory*, **80**, 94.

Athay, R. G. and Moreton, G. E., 1961, *Astrophys. J.* **133**, 935.

Babcock, H. W., 1961, *Astrophys. J.* **133**, 572.

Babcock, H. W., 1963, *Ann. Rev. Astron. Astrophys.* **1**, 41.

Babcock, H. W., and Babcock, H. D., 1955, *Astrophys. J.* **121**, 349.

Beckers, J. M., 1968, *Solar Phys.* **3**, 367.

Biermann, L., 1941, *Vierteljahrschrift Astron. Gesellschaft,* **76**, 194.

Billings, D. E., 1966, *A Guide to the Solar Corona*, Academic Press, New York.

Bray, R. J. and Loughhead, R. E., 1964, *Sunspots*, Chapman and Hall, London.

Bray, R. J. and Loughhead, R. E., 1967, *The Solar Granulation,* Chapman and Hall, London.

Bumba, V. and Howard, R., 1965, *Astrophys. J.* **141** 1492.

Chvojkova E., 1965, *Bull. Astron. Inst. Czech.* **16**, 57.

Cowling, T. G., 1946, *Mon. Not. Roy. Astron. Soc.* **106**, 218.

Cowling, T. G., 1957, *Magnetohydrodynamics*, Interscience Publishers, New York.

de Jager, C., 1965, in *Introduction to Solar Terrestrial Relations,* J. Ortner and H. Maseland, eds., D. Reidel, Dordrecht, Holland.

Ellison, M. A., 1963, *Quart. J. Roy. Astron. Soc.* **4**, 62.

Fichtel, C. E. and McDonald, F. B., 1967, *Ann. Rev. Astron. Astrophys.* **5**, 351.

Ginzburg, V. L., 1946, *Comptes Rendus Acad. Sci. U.S.S.R.* **52**, 487.

Goldberg, L., 1967, *Ann. Rev. Astron. Astrophys.* **5**, 279.

Grotrian, W., 1939, *Naturwiss.* **27**, 214.

Hess, W. N., ed., 1964, *AAS-NASA Symposium on the Physics of Solar Flares,* NASA SP-50, Washington, D.C.

Howard, R., 1967, *Ann. Rev. Astron. Astrophys.* **5**, 1.

Howard, R. and Babcock, H. W., 1960, *Astrophys. J.* **132**, 218.

Kippenhahn, R. and Schlüter, A., 1957, *Zeit. Astrof.* **43**, 36.

Kundu, M. R., 1965, *Solar Radio Astronomy*, Interscience Publishers, New York.

Leighton, R. B., 1964, *Astrophys. J.* **140**, 1120.

Leighton, R. B., 1968, *A Magneto-Kinematic Model of the Solar Cycle,* preprint.

Lüst, R., 1965, *Stellar and Solar Magnetic Fields*, North-Holland, Amsterdam.

Lüst, R. and Scholar, M., 1966, *Z. Naturforsch,* **212**, 1098.

Martyn, D. F., 1946, *Nature (London)*, **158**, 632.

Moreton, G. E., 1964, *Astron. J.* **69**, 145.

Newkirk, G., 1967, *Ann. Rev. Astron. Astrophys.* **5**, 213.

Oster, L., 1968, *Solar Phys.* **3**, 541.

Osterbrock, D. E., 1961, *Astrophys. J.* **134**, 347.

Parker, E. N., 1955, *Astrophys. J.* **122**, 293.

Parker, E. N., 1963, *Astrophys. J.* **138**, 552.

Parker, E. N., 1964, *Astrophys. J.* **140**, 1170.

Piddington, J. H., 1954, *Astrophys. J.* **119**, 531.

Piddington, J. H., 1958, *Phys. Rev.* **112**, 589.

Piddington, J. H. and Minnett, H. C., 1951, *Aust. J. Sci. Res.* **4**, 131.

Pikelner, S. B., 1965, *Vistas in Astronomy* **6**, 131.

Pikelner, S. B. and Livshitz, M. A., 1964, *Astron. Zh.* **41**, 1007.

Sakurai, K., 1967, *Rep. Ionospheric Res. Japan* **21**, 213.

Severny, A. B., 1964 *Space Sci. Rev.* **3** 451.

Shklovskii I. S., 1965, *Physics of the Solar Corona*, Pergamon Press, London.

Smith, H. J. and Smith, E. V. P., 1963, *Solar Flares*, Macmillan, New York.

Stickland, A. C., ed., 1968, *Annals of the IQSY*, MIT Press, Cambridge.

Svestka, Z., 1966, *Space Sci. Rev.* **5**, 388.

Svestka, Z., 1967, *I.A.U. 13th General Assembly, Prague, Agenda and Draft Reports*, 125.

Takakura, T., 1967, *Solar Phys.* **1**, 304.

Tandberg-Hanssen, E., 1967, *Solar Activity*, Blaisdell, Waltham, Mass.

Uchida, Y., 1968, *Solar Phys.* **4** 30.

Van Allen, J. A. and Krimigis, S. M., 1965, *J. Geophys. Res.* **70**, 5737.

Vostŕy, J., 1967, *Bull. Astron. Inst. Czech.* **18**, 37.

Wild, J. P., Sheridan, K. V. and Kai, K., 1968, *Nature (London)* **218**, 536.

Wild, J. P., Smerd, S. F. and Weiss, A. A., 1963, *Ann. Rev. Astron. Astrophys.* **1**, 291.

Xanthakis, J., ed., 1967, *Solar Physics*, Interscience Publishers, New York.

Zirin, H., 1966, *The Solar Atmosphere*, Blaisdell, Waltham, Mass.

CHAPTER 6

The Interplanetary Medium and Geomagnetic Cavity

The streaming of particles from the sun leads to a number of geophysical and astrophysical effects which have been studied since the beginning of the century. The aurora, geomagnetic activity, cosmic-ray variations and comet tail orientations all provide evidence of *solar corpuscular radiation* near the earth. Studies of these effects had, by the mid 1930's, provided a fairly strong case for such radiation.

This work had its origin in the ideas and terrella experiments of Birkeland and some suggestions by Lodge near the turn of the century. These led up to the theoretical works of Störmer and later of Chapman (for references, see Section 1.2). Störmer attempted to explain the aurora in terms of streams of individual (not interacting with one another) particles of energy several MeV, emitted from the sun. Although this explanation is now known to be incorrect, the work is the basis of studies of cosmic-ray trajectories. The idea of a *coherent stream* of particles leaving the sun is found in the discussion which led up to work of Chapman and Ferraro on the theory of geomagnetic field perturbations. Occasional blasts of electrically *neutral plasma* were envisaged, emitted at the times of flares. The one or two day delay of the geomagnetic storm after the flare indicated a velocity of propagation of about 10^3 km/sec. Later, Bartel's studies of recurrent magnetic storms and their association with hypothetical M-regions on the surface of the sun introduced the concept of a *continuous plasma stream* (Chapman and Bartels, 1940; Allen, 1944). Although continuous, the stream was localized and the flow past the earth was thought to be intermittent.

Another line of approach was that of Forbush, who showed that the intensity of the primary, or galactic cosmic rays varies over the 11-year cycle of solar activity. The cosmic-ray intensity tends to be higher when solar activity is low. Forbush also pointed out the tendency for the galactic cosmic-ray intensity to decrease during a geomagnetic

101

storm (for a review, see Forbush, 1954). Both of these effects might be explained by solar plasma moving past the earth, provided that the plasma carried a *magnetic field*. The possibility of an interplanetary magnetic field had been stressed by Alfvén in the early 1930's; such a field moving past the earth would also give rise to an electric field (discussed in Section 7.3).

It has long been known that the gaseous tails of comets point away from the sun irrespective of the direction of motion of the comet itself. It was generally thought that this effect is due to radiation pressure, but in 1951 Biermann showed that this could not be so and that again corpuscular radiation must be invoked (for a review, see Biermann in Chang and Huang, 1963). Also, since comets (of a suitable type) always showed this effect, it appeared that the solar corpuscular radiation was emitted in all directions and at all times, and not in a confined stream as thought previously.

The explanation of this flow was given by Parker in 1958, following Chapman's (1957) demonstration that the very high thermal conductivity of the solar corona must lead to a large outward flow of heat. Far from the sun the gas becomes too hot to be gravitationally contained, and streams away as indicated in Section 6.1.

In 1960 the first direct measurements of plasma flow were made from lunar and interplanetary space stations (Gringauz *et al.*, 1960). These observations were not very prolonged, but when the Mariner 2 results were reported (Neugebauer and Snyder, 1962) the existence of a solar wind was generally accepted. The first extensive direct measurements of the interplanetary magnetic field were made from the Pioneer 5 spacecraft (Coleman *et al.*, 1960). These results, like the solar wind observations, were in general agreement with the theory. Reviews of interplanetary plasma and fields have been given by Lüst and by Parker (in King and Newman, 1967), Dessler (1967), Ness (1968) and Wilcox (1968).

The third ingredient of the interplanetary medium is the *cosmic-ray population*, comprising a rather steady galactic, or primary, component and a highly variable solar component. The density of the former near the earth is determined by the inward diffusion and outward convection by the solar wind and magnetic field. The latter is highly variable in time and place in the solar atmosphere, and the resulting flux near the earth is then controlled by the field. These questions are discussed briefly in Section 6.3.

The configuration of the *geomagnetic field* under the influence of passing solar plasma has been a subject of some interest for several decades, but only relatively recently has the cavity been mapped by magnetometers and particle detectors on spacecraft. In addition to the cavity itself, a shock wave forms on the solar side and a region of compressed field and thermalized gas between the two. The region between the ionosphere and the cavity boundary is usually referred to as the *magnetosphere*, the boundary region is called the *magnetopause* and the region between this boundary and the shock is called the *magnetosheath*. These regions are discussed in Section 6.4.

Prior to the launching of spacecraft, the principal tool used in the study of the interplanetary plasma was the *geomagnetic activity*. The relationship between the wind strength and activity was not understood, but it was generally believed that a large increase in wind strength would give rise to increased activity and that perhaps the latter resulted mainly from rapid fluctuations in wind strength. When direct measurements were possible it was expected that the relationship would be quickly revealed, but this has not been the case. As seen in Section 6.5, the relationship is complex, and is not yet fully understood.

6.1 The Interplanetary Plasma

A mathematical development of the expanding solar atmosphere, or solar wind, has been given by Parker (1963, 1965) and papers referred to therein. Here we will briefly outline the physical processes involved.

The equation of motion of a spherically symmetric solar atmosphere under steady conditions depends on the gas pressure $N(r)kT(r)$, where N and T are the particle density and temperature, both functions of heliocentric distance r, and k is Boltzmann's constant. It also depends on the gravitational force $GM_\odot r^{-2}$ per unit mass, where G is the gravitational constant and $M_\odot$ is the solar mass. The pressure gradient and gravitational force combine to give the rate of change of momentum,

$$V\frac{dV}{dr} = -\frac{1}{NM}\frac{d}{dr}2NkT - \frac{GM_\odot}{r^2} \tag{6.1}$$

where V is the radial velocity and M is the particle mass. Mass flow for the steady state is constant, so that

$$NVr^2 = N_o V_o a^2 \tag{6.2}$$

where $r = a$ is a reference level where $N = N_o$, $V = V_o$. These equations cannot be solved unless some temperature distribution is found or assumed.

The very high thermal conductivity of the solar corona and interplanetary space (except across a magnetic field, it is $6 \times 10^{-7}\ T^{5/2}$ erg/cm/sec) must cause high temperatures to extend far into space.

Assuming a static model, the temperature falls off as $r^{-2/7}$ while the gravitational potential decreases as r^{-1}. Thus the thermal kinetic energy of each particle decreases with increasing r much more slowly than the energy needed to release the particle from gravitational control. One might expect then, that far enough from the sun the gas must be free to expand outwards and escape altogether from the sun, and this is the case. Solutions to equations 6.1 and 6.2 for the simple case of $T =$ constant show that the corona would expand at supersonic velocities of several hundred kilometers per second. Solutions with an outward decrease in temperature of the polytrope form,

$$T(r) = T_0 \left\{ \frac{N(r)}{N_0} \right\}^{\alpha - 1} \tag{6.3}$$

show that the expansion is not critically dependent on the form of $T(r)$, nor on the value of T_0. The theoretical velocity near the orbit of the earth lies in the range 200–2000 km/sec for coronal temperatures 0.5–4×10^6K. The predicted plasma density varies more rapidly with the temperature T_0 from $7/\text{cm}^3$ for $T_0 = 10^6$K to $100/\text{cm}^3$ for $T_0 = 2 \times 10^6$K.

The solar wind has been studied in detail by the spacecraft Mariner 2 and numerous others (Mackin and Neugebauer, 1966; Ness, 1968). The velocity is approximately radial, and near the earth's orbit its magnitude is in the range 300–400 km/sec in quiet times, increasing to 600–700 km/sec in disturbed intervals. Plasma densities generally lie within the range 1–20/cm³ and temperatures from about 10⁴K to a few times 10⁵K.

An inspection of the coronal irregularities sketched in Figure 1.1 suggests that there should be marked variations in the strength of the solar wind in different sectors around the sun, or in a given region at different times as the sun rotates. This sector structure is indeed the principal structural feature of the solar wind and of the interplanetary magnetic field. As Chapman and Ferraro showed, a single jet of plasma ejected from the sun would form a "garden hose" spiral, and the same is true for a continuous inhomogeneous coronal emission, so that a "sector" is a spiral segment of interplanetary space. Within a typical sector, shown in Figure 6.1, the plasma density and velocity show regular variations with both quantities a maximum shortly after the spacecraft enters the sector (Wilcox et al., 1967). The sector shown enveloped the spacecraft for 8 days; the solar rotation period is about 25 days, somewhat more when observed from spacecraft. The magnetic field varied from 3.5 to 6.5 gammas (1 $\gamma = 10^{-5}$ gauss), the wind velocity varied from 280 to 340 km/sec and the wind density varied from 7 to

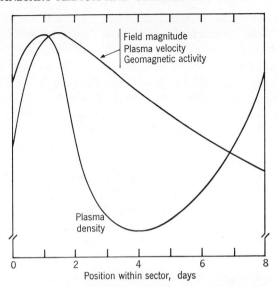

Fig. 6.1. A schematic representation of average interplanetary plasma (solar wind) density and velocity, magnetic field strength and geomagnetic activity as a function of position within the solar sectors. A sector is a spiral segment of interplanetary space containing an outward directed (or inward directed) magnetic field, shaped by solar wind flow combined with solar rotation.

$14/cm^3$. The geomagnetic activity, as measured by the 24 hour sum of the Kp index (see Section 6.5), varied from 10 to 25.

During periods of solar activity, and particularly following a flare, the interplanetary plasma density may increase to as much as $100/cm^3$ and the velocity may increase to more than 1000 km/sec; this is the interplanetary blast wave discussed in Section 5.4. These blast waves are responsible for most of the great geomagnetic storms and auroral activity, which are discussed in later chapters. A review of the enhanced solar wind and associated effects has been given by Obayashi (1967).

In addition to the interplanetary sector structure, there exist smaller-scale discontinuities in the interplanetary plasma, and these are important in connection with geomagnetic activity. A classical geomagnetic "storm" often has a sudden commencement (SC) which Chapman interpreted as the arrival of a discrete plasma cloud or stream. With the discovery of the solar wind this idea required modification, and Nishida (1964) suggested that SCs might be caused by tangential discontinuities propagating with the solar wind. Such discontinuities have been observed by Gosling *et al.* (1967) in a few cases which were apparently not related to flares. Physically the shear surfaces may be extensions of the boundary of coronal streamers (Figure 1.1) or other coronal features and may be part of a steady state (but rotating), inhomogeneous solar wind. Another type of solar wind discontinuity, proposed by Gold (1955) and others, is a collisionless shock wave propagating out from the sun. Although the particles suffer no collisions the medium has some fluid properties by virtue of the magnetic field (Section 2.4). A number of such shock waves have been observed by Gosling *et al.* (1968), with velocities significantly less than the mean transit velocity from the sun (derived from the time delay after the initiating flare). Evidently the shocks are decelerated on their way to the earth.

A review of interplanetary shocks and shears is given by Colburn and Sonett (1966) and their geomagnetic effects are discussed in Section 6.5. The effect of the solar wind in removing angular momentum from the sun (by magnetic stresses) has been discussed by Weber and Davis (1967) and in connection with stars in Section 11.3.

6.2 The Interplanetary Magnetic Field

Prior to the discovery of interplanetary plasma, this region was regarded more or less as a vacuum which might have been invaded by the solar general magnetic field or by the galactic field, depending on which would be strongest. Forbush's discovery of galactic cosmic-ray variations indicated a varying magnetic field which was capable, apparently, of sweeping some galactic radiation away from the earth. Morrison (1956) suggested a model comprising discrete clouds of solar plasma containing disordered magnetic fields, which were ejected at intervals from the solar atmosphere. The great flare event of February 1956 provided a variety of cosmic-ray data, both solar and galactic, of such profusion that most field models had to be abandoned because of inconsistency with one or other of the data. A configuration in the form of a magnetic loop or tongue, drawn out by ejected solar plasma, could explain the cosmic-ray data (Section 5.4), but during the ensuing months would be wound into a tight spiral. The continuous solar wind prevents this winding up because its average energy density (about 10^{-8} erg/cm^3) is very much greater than that of the field (about 10^{-10} erg/cm^3), and so it controls the field.

The combination of solar rotation and outward flow of plasma forces the frozen-in magnetic field to the form of an *Archimedes spiral*. The radial and azimuthal field components in the equatorial plane are given by

$$B_r(r) = B_o\left(\frac{a}{r}\right)^2 \qquad B_\varphi(r) = B_o\,\frac{a^2\Omega}{rV} \qquad (6.4)$$

where B_0 is the radial field at the base of the corona ($r = a$), V is again the wind velocity and Ω is the angular velocity of the sun. A wind velocity of 500 km/sec sets the field near the earth at an angle of about 45° to the radial direction. The sense of the field must, of course, correspond to that of the same field line traced back to the surface of the sun, and the spiral configuration means that a field of 1 gauss at the surface leads to about 3×10^{-5} gauss near the earth. The first experimental demonstration of this predicted spiral structure was given by McCracken (1962). Using data from ground-based neutron monitors he made a detailed analysis of the propagation of solar cosmic-ray particles and concluded that the flux is greatest from a direction 50° to the west of the sun and possibly about 10° north of the plane of the eliptic. Since the particles will spiral along the magnetic field lines, the latter make an angle of about 50° with respect to the sun-earth line, which accords well with the above model.

With the introduction of more sophisticated magnetometers on spacecraft, the direction and strength of the interplanetary field have been studied in great detail (Ness, 1968; Wilcox et al., 1967, Wilcox, 1968). The configuration is generally of the spiral form, with strength usually in the range 3–10 gammas.

The *sector structure* which rotates with the sun is most marked in its magnetic component because in adjacent sectors the field is alternately towards and away from the sun. A structure plotted by Ness and Wilcox (1965) is reproduced in Figure 6.2. This configuration is inferred from the magnetometer data received from the IMP-1 satellite during several solar rotations. For about 2/7 of the total circumference the spiral field is directed away from the sun, for the next 2/7 towards the sun, then 2/7 away and finally 1/7 towards the sun. The sector structure corotates with the sun, the field lines being rooted in the sun and also frozen into the solar wind. Although the latter moves radially, its streamlines assume the well-known "garden hose" spiral form, as shown by Chapman. The field lines must follow the streamlines and so form a spiral corotating structure. As seen from Figure 6.1, the preceding portions of the sectors are regions of maximum field magnitude, solar wind velocity and geomagnetic activity; these enhancements are shown schematically by the four hatched areas. The plasma density shows a minimum near the center of a sector and cosmic-ray intensity shows a minimum in the preceding portion.

The interplanetary field and plasma configurations are more complicated than this basic model in several ways.

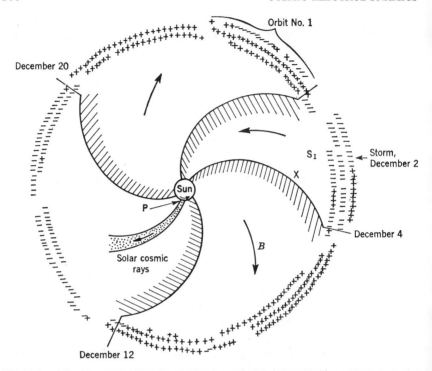

Fig. 6.2. The interplanetary sector structure determined from magnetometer measurements. The sector fields are predominantly away from the sun (+ signs) and towards the sun (− signs). The preceding part of a sector (hatched) tends to be a region of stronger magnetic field and larger solar wind velocity. Diffusion of solar cosmic rays from an area marked **P** as illustrated.

(1) Within each sector there are occasional reversals of the radial component, as indicated by the + (directed outwards) and − (inwards) signs of Figure 6.2, where each sign corresponds to a 3 hour average. Some at least of these reversals are interpreted as localized kinks in the spiral field as revealed by cosmic-ray data (Section 6.3). These bent force tubes are small scale features which are filamentary in form and appear to connect with particular regions in the solar atmosphere.

(2) The sector structure may remain quasi-stationary over a year or longer, during the declining years of the sunspot cycle. However, evolutionary changes do take place and some rather abrupt changes appeared with the advent of a new cycle in 1965. The altered field direction marked Orbit No. 1 in Figure 6.2 may have corresponded to a sector change.

(3) In addition to regular changes in solar wind velocity, as shown schematically in Figure 6.1, there are randomly occurring increases associated with flare events. On these occasions, one of which is marked Storm, December 2 in Figure 6.2, the sector involved is convected from the sun to the earth a day or two sooner than average. The sector boundary marked X arrived too soon on this occasion so that sector S_1 was invaded by outward directed magnetic field.

(4) The interplanetary field deviates from the ecliptic plane, and occasionally is directed approximately perpendicularly to this plane. These deviations are found to correlate strongly with geomagnetic activity, the latter being more notable for southward directed field, as described in Section 6.5.

(5) For some purposes it is desirable to describe the various random hydromagnetic disturbances in the interplanetary medium in terms of a power spectrum of fluctuations superimposed on the general spiral field and its sector structure. Such power spectra have been determined by Coleman (1967) and by Siscoe *et al.* (1968) who give references to earlier work. The latter use the data from Mariner 4, within the frequency range 3×10^{-4} to 0.5 cycle/sec, and find that the power dependence on frequency f has the form $f^{-\beta}$ with β typically near 3/2, and essentially always between 1 and 2 for each of the three components of the field. The results were obtained near solar minimum, and the power level was generally higher by a factor of about 10 on disturbed days compared with quiet days. These results are important in respect to cosmic-ray diffusion and are discussed in Section 6.3.

As we have seen in Section 4.2 magnetic field annihilation and reconnection may occur wherever fields of opposite sense are pressed together. This situation seems to be met in the boundary regions between adjacent sectors of the interplanetary field, and has been discussed in this connection by a number of writers including Gopasyuk and Křivský (1967) who use it in an attempt to explain some Forbush effects. Another aspect concerns the maintenance of a steady interplanetary field strength over periods of many years or decades. At the time of a flare event it seems likely that some fresh magnetic field lines are drawn out into interplanetary space (Section 5.4) and if this effect continues without a reverse flow, then the interplanetary field strength will grow continuously. Reverse flow would be provided by reconnection across sector boundaries in the manner shown in Figure 5.5 (which refers, of course, to a much lower level in the solar atmosphere).

6.3 Interplanetary Cosmic Rays

An important ingredient of the interplanetary medium is the high energy charged particle flux of galactic and solar origin. Particles interact with the magnetic field and give important information about its configuration, as we have seen above. The observational data are obtained from ground-level recorders, the most useful being the neutron monitor (for a review and discussion see Simpson, 1960) and also from recorders on spacecraft.

The variations in the intensity of galactic cosmic rays are the 11-year variation, the diurnal variation and the Forbush effect. As solar activity increases and decreases, galactic cosmic radiation shows an inverse variation (with a time lag). The variation depends on the latitude of the observing monitor and the threshold rigidities recorded, but is typically about 10%. The diurnal variation is small (about 0.2%) and irregular in amplitude. The direction of maximum intensity outside the geomagnetic field is about 90° east of the sun-earth line. The Forbush effect is a large intensity decrease (up to about 15%) over a period of some hours, followed by a recovery over a period of a week or so. This decrease is closely associated with magnetic storms and flare events.

In addition to these variations in galactic cosmic rays there are much larger variations of solar cosmic rays, the most striking being the *cosmic-ray flare* (McCracken, 1962; Obayashi, 1964). In this case the source of the particles is known and so some knowledge of the trajectory is available. The most important point is that there is a preponderance of westerly flares, as would be expected from the spiral field if the particles followed the field lines. Cosmic-ray events which do not originate in flares on the east side of the sun start several hours after the flare, the delay generally being greater than that for westerly flares. The propagation of solar cosmic rays differs from that of the plasma clouds ejected at the time of a flare event. The former comprise mainly protons with energies about 10^8 eV and so move much faster than the latter, whose proton energy is of order 10^3 eV. In addition, plasma clouds have the most notable geophysical effects when the flare event is near the center of the disk, and the propagation time does not depend on the position, east or west, on the disk. Evidently the plasma clouds control the interplanetary field while the field controls the cosmic rays. This is more evident when cosmic rays are injected while a plasma cloud (from a previous event) is in transit to the earth. The cosmic-ray time variation then reveals a complex feature: a very slow rise until the time of onset of the storm and then a very rapid rise. Clearly the plasma cloud has changed the field configuration and this has delayed the arrival of most particles until the cloud itself.

Solar cosmic rays are mainly protons in the energy range 10^6–10^9 eV, the flux decreasing with increasing energy from about $10^3/cm^2/sec$ to $1/cm^2/sec$ or less. These have been studied extensively, not only from spacecraft but by their ionizing effects in the ionosphere over the polar caps (Section 8.2), and occasionally at sea level. Van Allen and Krimigis (1965) have found that sub-relativistic electrons also escape into interplanetary space from flare regions. Their fluxes are strongly anisotropic, being collimated well to the west of the sun.

In developing a theory of cosmic-ray variations, we may start with those of relatively low energy, whose cyclotron radius is smaller than the smallest significant field kink. These particles travel along the field lines, and if they originate at a particular point P on the sun (Figure 6.2), then they will spread along the spiral field as shown by the dotted area. On this basis magnetic fine structure of filamentary form and dimensions about 10^5 km was recognized by McCracken and Ness (1966). Their model of twisting and overlapping field and cosmic-ray filaments is descriptively termed the "wet spaghetti" model. In general, the magnetic field strength decreases in the direction away from the sun and so, as shown in Section 8.4, the cosmic-ray trajectories become aligned more and more closely with the magnetic field lines. However, when an interplanetary blast wave is in transit between the sun and the earth the field is compressed and strengthened within the blast region. This "shell" of enhanced field acts as a barrier to some of the solar cosmic rays which are emitted behind the shell, and so their arrival at the earth is delayed. When the shell reaches the earth (storm onset) there is a very rapid increase in cosmic-ray intensity, and the subsequent decay is delayed because the blast wave continues to act as a barrier after it has passed the earth's orbit.

The picture of solar particles moving in simple helical paths along the lines of force must be modified when their energy is so great that their gyro (or cyclotron) radii are comparable with the magnetic irregularities. Particles move along the smooth sections of the lines of force and are deflected at the irregularities and it is the cumulative effect of a large number of such random deflections which results in a diffusion-like process. The combination of a spiral corotating magnetic field system and magnetic scattering centers which are carried out by the solar wind was suggested by Parker (1963). Far outside the earth's orbit, at distances perhaps between 10 and 100 AU, the solar wind is slowed by the interstellar medium and the spiral field becomes tangled and acts as a barrier to the galactic cosmic rays. The theory has been extended by Axford (1965) to take into account scattering centers between the barrier and the sun and hence of shorter term cosmic-ray fluctuations.

Suppose that the diffusion is isotropic and characterized by a mean free path λ, which will depend on the energy of the cosmic-ray particle. If there exists a gradient in the number density n of cosmic rays of velocity c, then the flux of particles at a given point due to this gradient is

$$\phi = -\frac{c}{3} \lambda \nabla n \tag{6.5}$$

from simple diffusion theory. This equation holds if the scattering centers are stationary; however, if they are moving with velocity $\mathbf{V}$ away from the sun there is an additional contribution caused by their sweeping effect, so that

$$\phi = \mathbf{V}n - \frac{c}{3} \lambda \nabla n \tag{6.6}$$

If there are no sinks or sources, then the equation of continuity is

$$\frac{\partial n}{\partial t} + \nabla.\phi = 0 \tag{6.7}$$

or

$$\frac{\partial n}{\partial t} + \nabla.(n\mathbf{V}) = \nabla.(D\nabla n) \tag{6.8}$$

where $D = c\lambda/3$ is the diffusion coefficient which may be a function of position. For slow changes in cosmic-ray density such as are expected for the 11-year cycle, $(\partial n/\partial t) \sim 0$ and the equation may be integrated. The results are in fair agreement with observational data.

Similar diffusion theory may be applied to variations in solar cosmic rays (Krimigis, 1965), but in this case time variations are all important and motions of the scattering centers may often be neglected so that equation 6.8 reduces to

$$\frac{\partial n}{\partial t} = \nabla.(D\nabla n) \tag{6.9}$$

Observed values of time variations have been used by various authors to determine values of the diffusion coefficient D. However, the quantity D itself may not be measured directly and must be related to the spectrum of magnetic irregularities discussed above (Jokippi, 1967, who gives earlier references). It is found that

$$D = v^2 T/6 \qquad (6.10)$$

where v is the particle velocity and T is a measure of the relaxation time in velocity space. This quantity is given by

$$T = \left(\frac{m}{q}\right)^2 \frac{2v}{\pi P(k_o)} \qquad (6.11)$$

where

$$k_o = \frac{qB}{mv} \qquad (6.12)$$

m and q being the particle mass and charge (emu), B is the mean magnetic field, and $P(k_o)$ is the power of magnetic irregularities at the wave number k_o (wavelength $2\pi r$, where r is the cyclotron radius).

An interesting application of this theory has been made by Nathan and Van Allen (1968) to explain the curious effect that 75 MeV solar protons and 40 keV electrons appear to have similar diffusion coefficients in spite of their difference in rigidity. The result may be explained if the power spectrum of the interplanetary magnetic field varies as f^{-2} in the range $2.7 \times 10^{-4} < f < 0.5$ cycles/sec. This corresponds to the extreme limit found by Siscoe et al. (Section 6.2) from analysis of magnetometer data, so that observation and theory are in reasonable agreement.

6.4 The Geomagnetic Cavity and the Bow Shock

In the absence of an interplanetary plasma, the earth's dipole magnetic field would extend indefinitely in all directions. However, since a stream of solar plasma is unable to penetrate the field appreciably it must compress the field, at least on the solar side, to form a cavity around which the plasma flows. This was one of the first problems in magnetohydrodynamics and was solved in an idealized form by Chapman and Ferraro. Nearly 30 years later, the surface of the cavity was observed by a magnetometer in Pioneer 5, and has been mapped by instruments in a number of spacecraft including Explorer 12 (Cahill and Amazeen, 1963). The first comprehensive mapping on both the sunward and night sides was performed in 1963–1964 by IMP-1 (Explorer 18) and reported by Ness et al. (1964).

Meanwhile, a number of attempts were made to construct a

theoretical model of the geomagnetic cavity. Mead (in Mackin and Neugebauer, 1966, where earlier references are given) has discussed the Newtonian approach to this problem, in which the interplanetary magnetic field is neglected, the magnetosphere is assumed completely closed and any frictional forces on the surface are neglected. The external pressure of the solar wind is $\eta NMV^2\cos^2\psi$, where N, M, V are the number density, particle mass and velocity of the wind, ψ is the angle between the velocity vector and the surface normal, and η is a constant between unity (if the plasma is thermalized on impact) and two (for specular reflection). Equating this to the internal pressure $B^2/8\pi$ we have

$$B^2 = 8\pi\eta NMV^2\cos^2\psi \qquad (6.13)$$

The value of B is not known until the cavity shape is known, and the shape is not known until B is determined. A first approximation is made by assuming that the surface field is twice the component of the undisturbed geomagnetic field tangential to the boundary. Successive approximations are then made to find the true surface. The results are in fair agreement with observations on the solar side of the magnetosphere, where the surface is nearly a hemisphere of radius typically $10\ R_E$ (earth radii or 6378 km). In the dark hemisphere, however, the

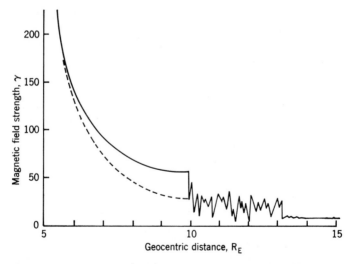

Fig. 6.3. A sketch of typical magnetic field intensity plotted against geocentric distance in a direction more or less towards the sun. The theoretical value of the (unconfined) geomagnetic field is shown as a dashed line. Beyond about 6 R_E the field is stronger than the theoretical value but drops abruptly at the magnetopause (at about 10 R_E) and again at the shock front (at about 13 R_E).

models fail because frictional interaction of the wind on the geomagnetic field draws the lines of force into a long tail (Chapter 7).

The existence of an interplanetary magnetic field endows the medium with the qualities of a fluid, and since the velocity past the geomagnetic cavity exceeds that of weak waves in the medium (which is the Alfvén velocity), a *bow shock wave* might form on the solar side of the cavity (Zhigulev, 1959; Axford, 1962; Kellogg, 1962). This idea is based on analogy with supersonic hydrodynamic flow and one would expect a shock of thickness a few times the proton cyclotron radius (about 1000 km) to form a few earth radii upstream from the magnetopause. Evidence of the bow shock was obtained by Freeman (1964) using electron flux data, and the magnetic field transitions in the shock and also in the magnetopause were demonstrated repeatedly by Ness *et al.* (1964).

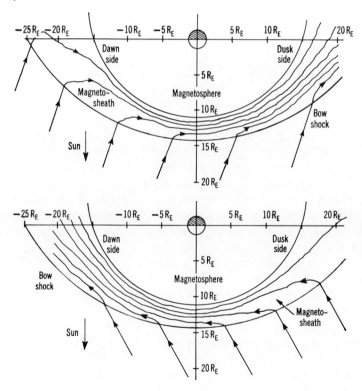

Fig. 6.4. The magnetosheath magnetic field (viewed from above the north pole) for the two possible orientations of the interplanetary field. In the top figure the interplanetary field is directed from the dawn to the dusk side of the earth (field direction may be reversed). In the bottom figure it is directed from dusk to dawn.

Thus magnetic fields in the neighborhood of the earth are of three distinct types, and these are illustrated in Figures 6.3 and 6.4. These refer to the solar side of the earth; on the anti-solar side there are further complications in the field structure.

Magnetosphere

The geomagnetic field itself extends to about $10R_E$, depending on the strength of the solar wind. The unconfined field would fall off approximately as R_E^{-3} (dashed curve) and the observed field does likewise to about 5 R_E, after which it falls off less rapidly but fairly smoothly. At the boundary there is an abrupt change in field strength and direction within a distance of a few hundred kilometers. The boundary or *magnetopause* moves inwards under the influence of an interplanetary blast (it has been observed at less than 7 R_E) and outwards during very quiet conditions. The physics of this region seem to be complicated and are discussed in Section 7.3.

Magnetosheath

The region between the magnetopause and bow shock is the magnetosheath and is a region of rather disordered field and thermalized, irregularly distributed plasma. The magnetic field strength in this region may vary as shown in Figure 6.3 (10–13 R_E) or it may be much more ordered, as stressed by Fairfield (1967). He finds that the field measured by IMP-1 and IMP-2 during 1963–1964 tended to be more ordered than the interplanetary field. The latter, with its irregularities, is convected through the shock, undergoing compression. The field lines are then draped around the magnetosphere in the manner shown in Figure 6.4. These sketches show the interplanetary and magnetosheath magnetic field lines projected onto the ecliptic plane. The field in the magnetosheath, between the magnetosphere and bow shock, may cross the sun–earth line from east to west or the reverse depending on the direction of the interplanetary field. The top orientation is more usual because of the average spiral configuration discussed above.

The magnetosheath field usually has a component perpendicular to the ecliptic plane, either northward or southward. In the flanks of the magnetosheath (on the dark side of the earth) Behannon (1968) has observed that the field generally has a large component perpendicular to the ecliptic plane, which alternates from positive to negative suggesting that filaments are observed that are alternately draped above and below the magnetosphere. This oscillation may have a period of a few hours and it is interesting to speculate on its possible origins: either the direction of the interplanetary field itself, or some twisting effect within the magnetosheath.

Bow Shock

The magnetic field of the bow shock has been mapped extensively by Behannon (1968, who gives earlier references). The shock is a well-defined surface of discontinuity, observable at distances as great as 75 R_E down the flanks. This is beyond the moon's orbit which is about 60 R_E. The plasma discontinuity has been described by Bridge *et al.* (1965) and others.

The existence of a bow shock was proposed in the first place by the analogy of supersonic flow past a solid obstruction. The required fluid properties of the medium are provided by the magnetic field and the shock is of the collision-free type, relying on non-linear interactions of waves to produce the randomization of particle motions. The plasma flow is slowed from supersonic (Mach number about 3–5) to subsonic, the bulk motion being converted to random motions. As the gas moves away from the shock it expands and is accelerated, eventually becoming supersonic again.

The mechanism of formation of a collisionless shock has been studied by Noerdlinger (1964) and others but is not fully understood. Scarf *et al.* (1965), who give earlier references, interpret the entire magnetosheath as a region of transition between interplanetary space and the magnetosphere. The experimental results leave no doubt that in general a sudden transition does occur, but on the other hand there are occasions when it is not evident or when it has very different forms to that shown in Figure 6.3. Sometimes there is a steady increase of field strength across the magnetosheath, on which are superimposed random oscillations. The bow shock, and its temporary absence, may play important parts in connection with geomagnetic activity; perhaps a well-defined shock provides a measure of protection for the magnetosphere from the solar wind and so reduces activity.

6.5 The Solar Wind and Geomagnetic Activity

Perturbations in the sea-level geomagnetic field were first connected with solar activity (flares) about a century ago and from this connection developed the concept of solar corpuscular radiation. Now that the solar wind has been studied extensively and correlated with geomagnetic activity over long periods, one would expect that the relationship between the two would be clear. However, apart from one simple form of activity, this is not the case.

A well-known perturbation is the sudden commencement (SC), distinguished by a rapid rise in the horizontal component of the surface field in equatorial and temperate latitudes. The rise time is typically 5 min and the rise may be preceded by a smaller fall; it is followed by a geomagnetic storm (Chapter 8). Similar perturbations which are not followed by storms are called sudden impulses (SI) and these may be positive or negative. All of these effects involve field perturbations of a few gamma to a few tens of gamma and they are simply explained in terms of sudden changes in solar wind pressure on the magnetosphere, according to equation 6.13. The change in B is transferred to the earth by a hydromagnetic wave. The source of the change is one or other of the two wind discontinuities discussed in Section 6.1.

Geomagnetic activity in general has a most complex morphology (Section 8.1), of which SCs and SIs comprise an almost negligible part. The enormous amount of data provided by numerous magnetometers distributed over the earth is too diverse to use in many correlation studies and so a general index of activity is required. For this reason the original K index was introduced by Bartels (see Lincoln, 1967) and later developed in the Kp index. The latter is a weighted average, for twelve observatories lying between middle latitudes and the auroral zone, of the intensities of fluctuation over each interval of 3 hours. Thus each Kp index is a single numerical (semi-logarithmic) measure of world-wide activity over a 3 hour period. By summing such indices a daily disturbance index is obtainable.

An examination of the contributions to Kp shows that the largest is provided by ionospheric current systems, concentrated in and near the auroral zones (Piddington, 1968). Also, as seen in Chapter 9, the main current systems appear to be driven by magnetospheric field lines as these move into and back from the geomagnetic tail. Thus the Kp index seems to provide a measure of the "stirring" of the magnetosphere by the solar wind. This phenomenon (or perhaps phenomena) is not understood, but it is not simply due to increased wind force, because some great magnetic storms develop with no SC.

Data from spacecraft have been used to study correlations between the interplanetary medium and Kp (Wilcox *et al.*, 1967; Fairfield, 1968, who give earlier references). There is no strong correlation between Kp and plasma density N, flux NV, kinetic pressure NMV^2 or temperature. However, there is a positive correlation with wind velocity and a most notable correlation with the southerly component of the interplanetary magnetic field and of the magnetosheath field. Fairfield uses a geomagnetic activity index AE which is similar to Kp except that the time interval is reduced from 3 hours to 2.5 min. This index is compared with the magnetosheath field and the results shown in Figure 6.5 where the number of cases with AE greater than the abscissa are plotted for four data groups. These groups are for weak or strong fields ($B \lesssim 15\ \gamma$) and for negative or positive directions (southward or northward fields). Large disturbances are found to correspond almost

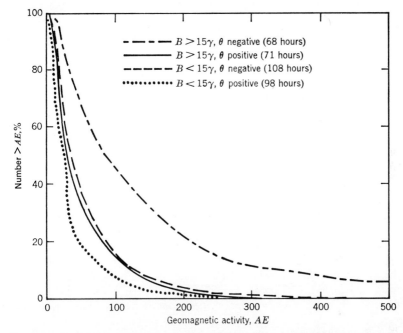

Fig. 6.5. Correlation of geomagnetic activity index (AE) with the strength and direction (θ negative is southward, θ positive is northward) of the magnetosheath field B. The ordinates are the percentages of cases with AE greater than the abscissa, for four groups as defined in the figure.

exclusively to large, southward fields. The possible significance of this result in connection with the formation of the geomagnetic tail is discussed in Chapter 7.

Some of the earlier evidence of a solar wind (confined to a narrow sector) is provided by recurrent geomagnetic disturbances with the 27-day solar rotation period. These were presumed to be caused by a long-lived solar stream originating on the surface of the sun and identified with a hypothetical M-region. Many statistical studies have been made in attempts to identify M-regions, but the conclusions vary. In view of the correlations discussed in Section 6.2 and illustrated in Figure 6.1, it would seem that the problem now reduces to relating features of the interplanetary sector structure (Figure 6.2) with features on the solar disk or in the corona.

Reviews of the interactions between the solar wind and the magnetosphere, for quiet and for disturbed conditions, have been given by a number of contributors in King and Newman (1967).

References

Allen, C. W., 1944, *Mon. Not. Roy. Astron. Soc.* **104**, 13.

Axford, W. I., 1962, *J. Geophys.. Res.* **67**, 3791.

Axford, W. I., 1965, *Planet. Space Sci.* **13**, 115.

Behannon, K. W., 1968, *J. Geophys. Res.* **73**, 907.

Bridge, H., Egidi, S., Lazarus, A., Lyon, E. and Jacobson, L., 1965, *Space Res.* **5**, 969.

Cahill, L. J. and Amazeen, P. G., 1963, *J. Geophys. Res.* **68**, 1835.

Chang, C. C. and Huang, S. S., eds., 1963, *Plasma Space Science Symposium*, D. Reidel, Dordrecht, Holland.

Chapman, S., 1957, *Smithsonian Pub.* **2**, No. 1.

Chapman, S. and Bartels, J., 1940, *Geomagnetism*, Clarendon Press, Oxford.

Colburn, D. S. and Sonett, C. P., 1966, *Space Sci. Rev.* **5**, 439.

Coleman, P. J., 1967, *Planet. Space Sci.* **15**, 439.

Coleman, P. J., Davis, L. and Sonett, C. P., 1960, *Phys. Rev. Letters* **5**, 43.

Dessler, A. J., 1967, *Rev. Geophys.* **5**, 1.

Fairfield, D. H., 1967, *J. Geophys. Res.* **72**, 5865.

Fairfield, D. H., 1968, *Goddard Space Flight Centre Rep.* X-612-67-338, and in press.

Forbush, S. E., 1954, *J. Geophys. Res.* **59**, 525.

Freeman, J. W., 1964, *J. Geophys. Res.* **69**, 1691.

Gold, T., 1955, *Gas Dynamics of Cosmic Clouds*, van de Hulst and Burgers, eds., North-Holland, Amsterdam.

Gopasyuk, S. and Křivský, L., 1967, *Bull. Astron. Inst. Czechoslovakia* **18**, 135.

Gosling, J. T., Asbridge, J. R., Bame, S. J., Hundhausen, A. J. and Strong, I. B., 1967, *J. Geophys. Res.* **72**, 3357.

Gosling, J. T., Asbridge, J. R., Bame, S. J., Hundhausen, A. J. and Strong, I. B., 1968, *J. Geophys. Res.* **73**, 43.

Gringauz, K. E., Bezruvkih, V. V., Ozerov, V. D. and Rubcinskii, R. E., 1960, *Soviet Phys. Doklady* **5**, 361.

Jokippi, J. R., 1967, *Astrophys. J.* **149**, 405.

Kellogg, P. J., 1962, *J. Geophys. Res.* **67**, 3805.

King, J. W., and Newman, W. S., eds., 1967, *Solar-Terrestrial Physics,* Academic Press. London.

Krimigis, S. M., 1965, *J. Geophys. Res.* **70**, 2943.

Lincoln, J. V., 1967, *Physics of Geomagnetic Phenomena*, S. Matsushita and W. H. Campbell, eds., Academic Press, New York.

Mackin, R. J. and Neugebauer, M., 1966, *The Solar Wind*, Jet Propulsion Laboratory, Pasadena.

McCracken, K. G., 1962, *J. Geophys. Res.* **67**, 447.

McCracken, K. G. and Ness, N. F., 1966, *J. Geophys. Res.* **71**, 3315.

Morrison, P., 1956, *Phys. Rev.* **101**, 1397.

Nathan, K. V. S. K. and Van Allen, J. A., 1968, *J. Geophys. Res.* **73**, 163.

Ness, N. F., 1968, in *Annals of the IQSY*, A. C. Stickland, ed., MIT Press, Cambridge; *Ann. Rev. Astron. Astrophys.* **6**, 79.

Ness, N. F., Scearce, C. S. and Seek, J. B., 1964, *J. Geophys. Res.* **69**, 3531.

Ness, N. F. and Wilcox, J. M., 1965, *Science* **148**, 1592.

Neugebauer, M. and Snyder, C. W., 1962, *Science* **138**, 1095.

Nishida, A., 1964, *Rep. Ionosh. Space Res. Japan* **18**, 295.

Noerdlinger, P. D., 1964, *J. Geophys. Res.* **69**, 369.

Obayashi, T., 1964, *Space Sci. Rev.* **3**, 79.

Obayashi, T., 1967, *Solar-Terrestrial Physics,* King and Newman, eds., Academic Press, London.

Parker, E. N., 1963, *Interplanetary Dynamical Processes,* Interscience Publishers, New York.

Parker, E. N., 1965, *Space Sci. Rev.* **4**, 666.

Piddington, J. H., 1968, *Geophys. J. Roy. Astron. Soc.* **15**, 39.

Scarf, F. L., Crook, G. M. and Fredericks, R. W., 1965, *J. Geophys. Res.* **70**, 3045.

Simpson, J. A., 1960, *Astrophys. J. Suppl.* **4**, 378.

Siscoe, G. L., Davis, L., Coleman, P. J., Smith, E. J. and Jones, D. E., 1968, *J. Geophys. Res.* **73**, 61.

Van Allen, J. A. and Krimigis, S. M., 1965, *J. Geophys. Res.* **70**, 5737.

Weber, E. J. and Davis, L., 1967, *Astrophys. J.* **148**, 217.

Wilcox, J. M., 1968, *Space Sci. Rev.* **8**, 258.

Wilcox, J. M., Schatten, K. H. and Ness, N. F., 1967, *J. Geophys. Res.* **72**, 19.

Zhigulev, V. N., 1959, *Soviet Phys. Doklady* **4**, 514.

CHAPTER 7

The Earth's Magnetosphere and Tail

The model of the geomagnetic cavity discussed in the preceding chapter was based on the assumption that the external plasma would exert a pressure normal to the surface, and would be completely excluded from the cavity. A combination of observational data and theory has since revealed two additional features which enormously complicate the model. The first is a belt or belts of energetic charged particles trapped inside the cavity or magnetosphere. This *radiation belt* or Van Allen belt is a permanent part of the magnetosphere structure, extending deep into the magnetosphere until it is limited by the absorption of particles into the earth's atmosphere. The charged particles gyrate around the magnetic field lines and spiral along the lines from hemisphere to hemisphere, being reflected when the field strength is great enough. Particles close enough to the earth also drift around the earth, as a result of the inward gradient in field strength. The intensity and spectrum of this corpuscular radiation varies greatly with place and time, particularly during periods of magnetic activity.

As we shall see, the radiation belt is responsible for the main phase of a geomagnetic storm and indeed this cause and effect had been predicted on theoretical grounds shortly before its discovery (see Section 8.5). The belt also supplies some *auroral particles*, which are said to be "precipitated" into the atmosphere. The theory and observations of the mechanisms of origin and loss of the belt particles has constituted one of the major efforts in space physics, and is discussed in Section 7.1.

The model of the geomagnetic cavity discussed in the preceding chapter involves no dissipation of energy by the plasma stream, and as a consequence contains no tangential stress at the magnetosphere boundary. A drastic modification to this model may result from such a *frictional effect*; if strong enough, magnetic field lines will be carried from the front and sides of the magnetosphere and drawn out into a *long magnetic tail* (Piddington, 1959, 1960). Such a tail would consist of two sections, each comprising a bundle of magnetic force lines emerging from a polar cap. The field lines emerging from the southern hemisphere would be directed away from the earth in the antisolar direction and those from the northern cap would be directed towards the earth. These two bundles of field lines would tend to expand under the influence of magnetic pressure, but this would be resisted by pressure of the solar wind. Near the geomagnetic equatorial plane the two bundles would be pressed together to provide a *magnetic neutral sheet* separating regions of oppositely directed fields.

Just such a tail and its magnetic neutral sheet were mapped in 1965 by Ness (1965), using the IMP-1 (Explorer 18) Satellite. Some prior evidence of its existence was obtained in 1961 by Explorer 10 (Heppner *et al.*, 1963) and again in 1962 by Explorer 14 (Cahill, 1966). More recently measurements in the tail have been made at distances near the moon's orbit by the lunar-orbiting satellite Explorer 35 (Ness, Behannon *et al.*, 1967) and it has been observed in a somewhat weakened form at a distance of 1000 R_E, or about 6×10^6 km, by the space probe Pioneer 7 (Ness, Scearce *et al.*, 1967; Fairfield, 1968). It has been extensively mapped by the satellites IMP-1 (Ness *et al.*, 1966) out to $30R_E$, and more recently by IMP-2 or Explorer 21 (Fairfield and Ness, 1967) out to 16 R_E and by Explorer 33 out to 80 R_E (Behannon, 1968; Mihalov *et al.*, 1968). These results are discussed below.

In the original, theoretical model of the geomagnetic tail, the possibility of magnetic field annihilation and reconnection across the neutral sheet was considered, and it was suggested that the neutral sheet might be the source of energetic auroral and Van Allen particles. The neutral sheet has been mapped magnetically by Speiser and Ness (1967) and others who find some evidence of reconnection occurring. The plasma associated with the neutral sheet has been observed by Bame *et al* (1967) and others who find substantial fluxes of energetic particles.

A schematic representation of the whole magnetosphere and of the bow shock is shown in Figure 7.1. This is a section containing the earth's dipole NS and the sun–earth line. The geomagnetic field and magnetosheath field are delineated by a few field lines, but the boundaries are not shown. Magnetic field lines which close near the earth tend to be frozen into the ionosphere and to rotate with the ionosphere, and hence with the earth. For this reason the inner part of the field configuration is often called the *corotating magnetosphere*, as opposed to the *magnetotail* which is frozen into the solar wind and so always points in the sun–earth direction. The northern tail section contains field lines A, B and C which connect to a polar cap. As we shall see,

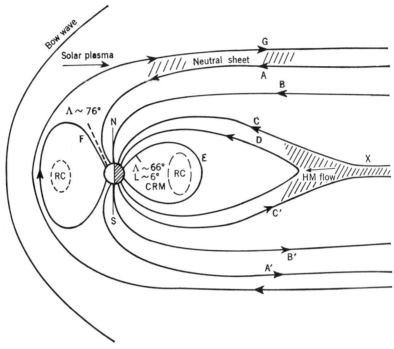

Fig. 7.1. A section of the magnetosphere containing the earth's dipole NS and the sun–earth line. Magnetic field lines A, B, C in the northern section and A′, B′, C′ in the southern section have been drawn out to form a magnetospheric tail. Field lines E and F are the outermost of the corotating magnetosphere (CRM), the boundary is at magnetic or invariant latitude $\Lambda \sim 76°$ near noon and $\Lambda \sim 67°$ near midnight. A magnetic neutral sheet X allows field-line reconnection, and line D is in transit from the tail back to CRM. A ring current, or plasma belt (RC), is shown in section. Field line G is in the magnetosheath and is part of neutral sheets lying immediately outside the tail.

this cap is not symmetrical about the pole but extends down to magnetic latitude $\Lambda \sim 76°$ near the noon meridian and $\Lambda \sim 67°$ near the midnight meridian. Corresponding field lines A′, B′ and C′ connect to the southern polar cap and lines C and C′ define the magnetic neutral sheet X. The corotating magnetosphere is defined by lines E and F (which lie near its boundary) and line D is in the process of contracting and moving from the tail to CRM. Field line G is an interplanetary, or rather magnetosheath, line draped over the cavity and corresponding to a northward directed interplanetary field. It is seen that other neutral sheets are formed above and below the tail; these are not present for a southward directed field but are replaced by a neutral sheet around the subsolar point. Finally a radiation belt and ring current (RC) or region of large flux of energetic, trapped particles is shown in section.

Geomagnetic tail models have been discussed by many authors, the main differences between models being the mechanisms by which field lines enter and leave the tail. The former mechanism must operate over the solar side of the magnetosphere and is discussed in Section 7.3. The latter operates in the tail neutral sheet, and perhaps the neutral sheet between field lines A and G (and A' and G) of Figure 7.1; it is discussed in Section 7.4.

7.1 Particles in the Corotating Magnetosphere and Tail

Some of the experiments carried by the early U.S. and U.S.S.R. earth satellites were designed to study cosmic rays and auroral and solar corpuscular radiation. The first report of the existence of geo-magnetically trapped radiation was based, curiously enough, on anomalously low counting rates of shielded geiger tubes in Explorers 1 and 3 when they reached altitudes above 1000 km. The rates fell below that of the primary cosmic radiation, and was correctly inter-preted as being due to saturation of the circuitry by high fluxes of particles. Because the radiation was confined to high altitudes, it was concluded that it was geomagnetically trapped and hence consisted of charged particles (Van Allen et al., 1958). It seems likely that the particles responsible were mainly penetrating protons. The geiger tubes on Sputnik 2 showed enhanced counting rates at high latitudes and lower altitudes (Vernov et al., 1958), which may have been caused by bremsstrahlung from trapped or precipitating electrons.

Reviews of the first four or five years of investigations of the earth's particles have been given by Van Allen (1963) and O'Brien (1963). Two regions of enhanced corpuscular radiation were recognized, their centers having geocentric distances (in the equatorial plane) of about $1.5 R_E$ and $3.5 R_E$. The flux and energy of the particles in these zones were found to vary, particularly during periods of geomagnetic activity. The changes were greatest for more distant regions, and beyond about $2 R_E$ were catastrophic for great storms. The first measurements made from a satellite of particles precipitating into the earth's atmosphere (Krasovskii et al., 1961) were consistent with earlier ground-based measurements of Doppler-shifted $H\alpha$ in auroral spectra (Meinel, 1950), with rocket observations of auroral zone x rays (Van Allen, 1957), and with auroral theory in general. For some time it was not clear just what were the particles making up the Van Allen belts, but it gradually emerged that both electrons and protons were present in roughly comparable intensities and energies down to about 40 keV for electrons and 100 keV for protons. It also appeared that most of the geomagnetic cavity is occupied by trapped particles and the terms inner and outer radiation zones have little meaning, although they are still used for convenience.

A second generation of magnetospheric particle experiments was responsible for many further advances which have been reviewed by various contributors in McCormac (1966). Previously the radiation zone had been regarded as a doughnut-shaped region, symmetrical about the geomagnetic dipole axis and free from diurnal control by the sun. As shown by O'Brien, however, the outer or high-latitude termination (boundary) of 40 keV electrons is at magnetic latitude about 75° near noon and about 69° near midnight. This effect is a result of the asymmetry of the magnetic field shown in Figure 7.1 and is discussed further below. A further complication, this one man-made, is the injection of charged particles into the magnetosphere by detonation of nuclear weapons. During the fissioning of uranium or plutonium, electrons of average energy about 1 MeV (up to about 8 MeV) are released at the rate of about 6×10^{23} electrons per kiloton of fission yield. Each explosion is able to produce a radiation belt and some half-dozen of these belts have been observed in some detail. The most notable resulted from the U.S. Starfish explosion on July 9, 1962, which extended from about 2 R_E down to the ionosphere. Study of these belts has yielded important information concerning the loss mechanism for particles trapped in different regions.

The main features of the natural and Starfish radiation zones are shown in Figure 7.2, which is a cross section of the magnetosphere perpendicular to the sun-earth line. On the right are the protons, with energies extending up to about 1 BeV for those near the earth, and having a maximum intensity for all protons above 40 MeV of about $10^4/cm^2/sec$ at about 1.5 R_E. Further out, low-energy protons (down to 1 MeV and less) are found to about 7 R_E, with a flux of about 10^7 protons/cm^2/sec above 100 keV. This flux corresponds to an energy density only about 0.1 that of the field, but as seen below lower energy protons sometimes provide energy comparable with that of the field and so have large geomagnetic effects. Throughout the trapping region there is a fairly systematic change of proton energy with geocentric distance, the more energetic protons occurring at smaller distances. Also shown in the figure is the motion of a trapped particle along field line 3 R_E (or $L = 3$); the particle spirals from hemisphere to hemisphere in the manner discussed in Section 8.4.

On the left-hand side of Figure 7.2 is shown the remnant of the Starfish belt five years after the explosion; electrons had a flux (for energies

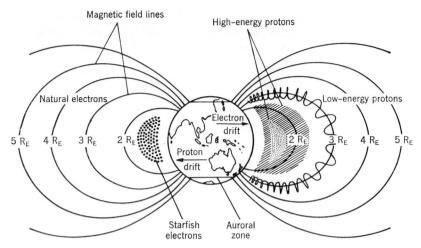

Fig. 7.2. The principal features of the geomagnetically trapped corpuscular radiation in a section containing the dipole, and perpendicular to the sun–earth line. The proton distribution is shown on the right and the electron distribution on the left, the latter including the results of a nuclear explosion (Starfish).

up to a few MeV) of about $10^7/cm^2/sec$ in mid-1966. Lower-energy electrons of natural origin are present throughout the whole region of ordered magnetic field, the average flux for energies above 40 keV is about $10^7/cm^2/sec$. The systematic variation with geocentric distance of the energies of these high-energy electrons is somewhat similar to that of the protons. So also is the flux and spectrum of the electrons, those at greater distances being subject to greater changes, particularly during periods of notable magnetic activity.

More recent measurements of the radiation belts have extended the spectrum downwards to provide fluxes of "low-energy" particles which Frank (1968) defines as protons below 100 keV and electrons below 50 keV. These measurements are more difficult, but electrons of energies about 1 keV and less have been measured in the magnetosheath, also just inside the corotating magnetosphere on the day side and just outside the corotating magnetosphere on the night side. Frank has also measured the energy density of low-energy protons on particular occasions within the magnetosphere. These provide evidence of a storm-time radiation belt and its ring current centered near $4\ R_E$, which might be capable of providing substantial geomagnetic effects as described in Section 9.1.

Yet another particle population of the magnetosphere is provided by thermal protons and electrons which are an extension of the ionosphere. There is some evidence, based largely on the interpretation of the propagation of radio "whistlers" through the magnetosphere, that this plasmasphere extends outwards to the field lines marked 4 R_E in Figure 7.2 (Carpenter, 1966). Beyond this magnetic shell the density falls quickly from about $100/cm^3$ to about $1/cm^3$.

The magnetotail is also filled with a complex distribution of energetic particles and hot plasma. The latter is in the form of a plasma sheet, in the center of which lies the magnetic neutral sheet. However, the plasma sheet extends north and south far beyond the neutral sheet; at a geocentric distance of 17 R_E its thickness is about 5 R_E near the center of the tail and 10 R_E near the edges. Thus the plasma sheet is roughly 25 times the thickness of the neutral sheet (Bame et al., 1967, who give earlier references).

Within the plasma sheet are more localized "islands" of electrons of energies up to about 200 keV (Anderson and Ness, 1966; Murayama and Simpson, 1968, who give earlier references). These islands seem to have linear dimensions about 1 R_E and are scattered throughout the plasma sheet. Within the magnetic neutral sheet itself there is a component of electrons of energies ranging above 200 keV. These are present continuously out to at least 30 R_E, but the flux is highly variable. Possible (separate) origins of the island electrons and neutral sheet electrons are discussed in Sections 7.4 and 9.2–9.)

The principal problems of the radiation belt and tail plasma sheet and islands are the source and loss mechanisms. The first *source* mechanism to be worked out was the decay of cosmic-ray albedo neutrons. Cosmic rays striking atmospheric nitrogen or oxygen nuclei cause the emission of neutrons, some of which scatter into the magnetosphere where their decay products, a proton and an electron per neutron, may be trapped. There is no doubt that such a source exists, but it cannot explain the large increases observed during magnetically disturbed periods and, if important at all, may explain only the high-energy protons of the inner belt. It is fairly certain that the main source is some particle acceleration mechanism or mechanisms operating in the tail and also in the corotating magnetosphere. This source is discussed in Sections 8.5, 9.1 and 9.2.

The one certain *loss* process is by collisions of the trapped energetic particles with atmospheric particles. Studies of the decay of radiation injected by high-altitude nuclear explosions, and following impulsive increases of natural origin, show that in the region out to 1.25 R_E this is the main loss mechanism. Beyond this magnetic shell some other and more powerful loss mechanism must exist. In the case of low-energy protons a loss occurs as a result of charge-exchange collisions, during which the proton acquires an electron and so escapes, leaving a charged particle with only thermal energy. The search for loss processes other than by particle collisions has been pursued vigorously for many years and is discussed in Section 8.4 in a general way, and in Chapter 9 in connection with a magnetospheric model proposed more recently.

7.2 Mapping the Magnetotail

The earth's magnetic tail is a permanent extension of the geomagnetosphere, extending far beyond the lunar orbit (about 60 R_E) and observed during one period at 1000 R_E. During geomagnetically quiet periods ($Kp \gtrsim 2 +$) field strengths in the tail vary from a low of about 4 γ to a maximum of about 40 γ, with values usually in the range 10–20 γ. The strength decreases with increasing geocentric distance, the upper value (20 γ) corresponding to distances about 10–20 R_E and the lower value to distances near 60 R_E. The field strength generally tends to increase during periods of geomagnetic activity. The tail comprises a north section (field directed towards the earth) and a south section (field away from the earth) as shown in Figure 7.1. The average radius of a section is about 20 R_E, but there is evidence of expansion and contraction, say from 15 R_E to 30 R_E, although it is difficult to separate this from a flapping motion which also occurs in response to changing wind direction and turbulence.

The important question concerning the magnetotail is the variation of magnetic flux into each section—that is, the changes with time and with geocentric distance of the average field strength in a section multiplied by the cross sectional area of the section. We will attempt to evaluate these changes and will then consider their causes.

The observed decrease in tail field strength with geocentric distance is interpreted as a result of some field lines extending only a limited distance and then connecting across the magnetic neutral sheet. The conservation of magnetic flux ($\nabla.\mathbf{B} = 0$) requires an average field of strength 2 γ crossing the sheet between 10 R_E and 80 R_E, so that the "neutral" sheet is only approximately neutral. A magnetic force tube which has connected in this manner will provide a tension $B^2/8\pi$ tending to make it contract, and unless this is balanced by a pressure gradient the force tube will contract until it rejoins the corotating magnetosphere (as line D and, later, line E of Figure 7.1). There will also be another loop of the reconnected force tube, lying beyond the point of reconnection and having a southward directed field in the neutral sheet. This loop or bubble will contract away from the earth and be lost.

These considerations suggest that magnetic field lines leave the tail, thereby decreasing its magnetic flux. In order to maintain the tail it is necessary, therefore, to have a mechanism of replenishment. Furthermore, there is evidence that the tail flux sometimes increases above its average value, which implies an accelerated flow to the tail (or perhaps a retarded flow from the tail). The observed increase of tail field strength with geomagnetic activity provides some evidence of increased flux, but it is difficult to separate this effect from tail compression by an enhanced solar wind. The most conclusive evidence of changing tail flux is provided by observations near the ionosphere where field compression is negligible. These results are also of interest in connection with auroral and geomagnetic activity, and so are discussed here in some detail.

Magnetotail lines such as A, B, C and A', B', C' in Figure 7.1 connect to the earth through the polar ionospheres, while all other lines connect at lower latitudes. It follows that in each hemisphere there must be a boundary, more or less circular and centered near the magnetic pole,

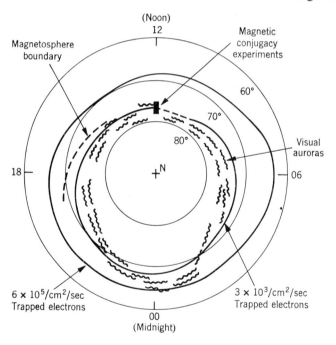

Fig. 7.3. A northern hemisphere plot in local-time magnetic coordinates (down to latitude 60°) of some ionospheric phenomena, and magnetospheric phenomena projected along field lines into the ionosphere. Full lines denote the median intensities of trapped electrons of energies above 40 keV. Wavy lines trace the auroral oval, and the rectangle denotes a region of changing magnetic conjugacy.

above which all field lines enter the tail and below which all field lines traverse the corotating magnetosphere. One would expect these boundaries to have considerable significance in connection with trapping of particles and perhaps other effects; only particles within CRM are trapped, those in the tail have an exit along the field lines. This is indeed the case, and some such phenomena are illustrated in Figures 7.3 and 7.4. These are schematic representations of the northern

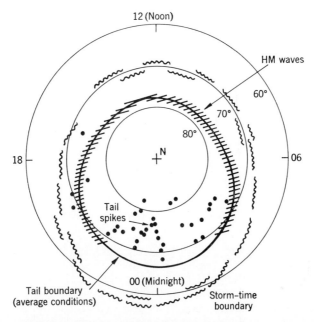

Fig. 7.4. A plot in the same coordinates as Figure 7.3, showing the magnetotail boundary (full line), observed strong hydromagnetic waves (cross hatching) and tail "spikes" of energetic electrons. The wavy lines delineate the greatly expanded auroral oval during a great magnetic storm.

ionosphere viewed from above the magnetic pole N and extending down to latitude 60°. They show plots of several phenomena, the first being that of auroras during average conditions of magnetic activity. The wavy lines of Figure 7.3 represent discrete, *visual auroras* (as opposed to diffuse auroras, usually detected by instruments), which occur around the "auroral oval." This is not centered at the magnetic pole but rather about 5° down the midnight meridian (Feldstein and Starkov, 1967, who give Feldstein's earlier references). This oval replaces the classical *auroral zone* which was approximately the 67° parallel of magnetic latitude. Such a zone had no real physical significance, but was inferred from observations made only during the night, when the auroral oval touched the 67° parallel.

The two full curves of Figure 7.3 summarize the data of Frank et al. (1964) for energetic trapped electrons seen above the ionosphere. There is a rapid decrease from a near maximum flux of 6 × 10^5/cm²/sec to a value lower by a factor of about 200, and this zone is regarded as a *trapping "boundary."* One would expect this boundary to coincide more or less with that between the corotating magnetosphere and the tail, since electrons in the tail would soon be lost. Further evidence of the tail boundary is provided by magnetic conjugacy experiments, made at two stations in opposite hemispheres and connected by a magnetic field line. Stations at latitudes corresponding to the low-latitude end of the small rectangle of Figure 7.3 record almost identical patterns of auroral and geomagnetic effects (near noon), thus indicating magnetic correction. Stations at the high-latitude end record different patterns indicating magnetic connection to the tail. One infers that the tail boundary intersects the rectangle and perhaps extends as the dashed line. All of these results correspond to conditions of average magnetic disturbance and are summarized by the full curve of Figure 7.4. Here further boundary evidence is provided by cross hatching, which represents the observed incidence of magnetohydrodynamic waves at a low-flying satellite (Zmuda et al., 1967). Finally the "spikes" of energetic electrons, also seen at low altitudes (McDiarmid and Burrows, 1965), are recognized as a tail plasma sheet phenomenon and so help to define the tail section projected into the ionosphere along the magnetic field lines. We conclude that the tail boundary for average conditions (*Kp* about 1–3) is defined by the full curve, with magnetic latitudes about 76° and 67° at noon and midnight, respectively.

The magnetic flux through this section, based on the sea-level field strength of about 0.6 gauss and the area of the auroral oval, is about 7 × 10^{16} gauss cm². This must also be the flux into the tail section, out to distances where it has been depleted appreciably by reconnection through the neutral sheet.

During a great magnetic storm the auroral oval is observed to expand to dimensions indicated by the wavy lines of Figure 7.4 (Akasofu, 1966). Simultaneously the high-latitude boundary of trapped energetic particles, and also the ionospheric current system associated with auroras, move to lower latitudes over a period of about 8 hours. This motion is complicated by a repeated widening of the band of visible auroras (characteristic period about 1 hour), as well as its bodily motion. However, at times the oval is clearly defined at latitudes below 60° magnetic as shown, and this and the associated data discussed in more detail in Chapter 9 provide strong evidence of a large increase of magnetic flux into the tail. The magnetic field strength in the ionosphere only changes by one or two percent during the largest magnetic storm, and so the tail flux must be indicated approximately by the area of the auroral oval. On this basis it will attain or exceed 10^{17} gauss cm^2 during great storms.

A tail with a flux of 10^{17} gauss cm^2 per section and dimensions 40 R$_E$ has a field strength of about 30 γ near the earth, in reasonable agreement with observed storm-time values. If this field extended to 100 R$_E$ it would contain magnetic energy of about 10^{23} erg, and as the characteristic period of a storm is 8 hours its energy requirement is a few times 10^{18} erg/sec, which is a few percent of the energy of the solar wind intercepted by the magnetosphere. This simple argument shows that a great deal of energy seems to be extracted from the solar wind to form the storm-time magnetotail and, as we shall see, this energy source is essential in explaining the storm-time radiation belt and auroral energy expenditure. However, the argument may greatly underestimate the total energy available because it neglects the effects of reconnection of field lines across the tail neutral sheet. Tail field lines have been observed beyond 100 R$_E$ behind the earth, although the magnetic flux there is apparently only a small fraction of that near the earth because of reconnection. Furthermore, the time taken for a field line to be drawn out to a length of 1000 R$_E$ by a wind of velocity 500 km/sec is only about half the characteristic storm time. These considerations suggest the possibility that all field lines are drawn out to this length and that the storm energy provided to the tail exceeds 10^{24} erg, although the magnetic energy in the tail at any given time is much less because of losses following reconnection.

An objection has been made to this large energy transfer because it is comparable with the total energy of the solar wind intercepted at the front of the magnetosphere. This objection is not valid, because as a line or tube of force is being extended work is done on it continuously, even at a geocentric distance greater than 1000 R$_E$. The frictional interaction responsible for the energy transfer must operate over the whole surface of the tail which is an area 10 or 10^2 times larger than the area seen when looking from the sun.

The actual form of the frictional interaction is discussed in the following section.

7.3 Physics of the Geomagnetic Boundary

In the preceding section it was shown that magnetic field lines are regularly moved from the corotating magnetosphere into the tail, where some at least are stretched out to a length exceeding 1000 R_E. This process seems to be greatly accelerated during magnetic storms, and since these correlate closely with a southward component of magnetosheath field (Figure 6.5) it seems likely that a southward field speeds up the process of moving lines of force into the tail. However, we should not overlook the other possibility, that tail growth and some resulting magnetic disturbance is caused by a temporary inhibition of reconnection across the tail neutral sheet.

An understanding of the process of moving field lines from the front or sides of the geomagnetic cavity into the tail requires study of the various effects which may occur at the boundary or magnetopause. One effect, suggested by Dungey (1961, 1967), is possible only when the field outside the cavity is directed southward. Since the geomagnetic field is directed northward, the two then form a magnetic neutral sheet and it is possible then for geomagnetic field lines to break and connect across the neutral sheet with interplanetary lines as discussed in Section 4.2. Some evidence for this, in the form of field lines crossing the boundary, has been given by Sonnerup and Cahill (1968). This effect also provides an immediate explanation of the observed correlation of a southward field and geomagnetic activity. However, there appear to be a number of alternative explanations of this correlation and these will be discussed.

The geomagnetic boundary of magnetopause was first studied by

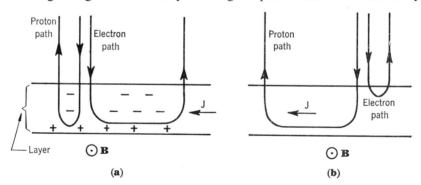

(a)   (b)

Fig. 7.5. Schematic representations of two models of a boundary layer. A plasma stream is directed from the top of the figure, downwards onto a magnetic field domain. The field **B** is outward from the paper and its strength falls to zero as we move upwards across the boundary layer. **(a)** A space-charge field is provided by the impinging protons, which penetrate a little deeper than the electrons, The latter are accelerated and so provide the terminating current sheet J. **(b)** The above proton space-charge field is neutralized by electrons sucked up from the ionosphere. The impinging electrons are now bounced off the boundary layer and the current J is carried by the penetrating protons.

Chapman and Ferraro and in more detail by Ferraro (1952), whose model is shown in Figure 7.5a. A cold plasma stream is incident from above on a magnetic domain limited by a boundary layer perpendicular to the stream. An individual proton would move under the influence of the Lorentz force $e\mathbf{v} \times \mathbf{B}$ and be reflected; because of its much greater momentum it would penetrate much further than an electron. In a plasma stream, however, the (upward) electric field resulting from this large separation of protons and electrons would be very large, and what happens is that a much smaller field is created as the protons and electrons penetrate to almost the same depths. This electric field is mainly responsible for slowing and reflecting the protons. It also accelerates the electrons, which gain energy nearly equal to that of the protons outside the layer. The electrons are reflected by the magnetic field after moving a considerable distance along the layer. This motion provides the current sheet J which terminates the magnetic field **B** according to Maxwell's equation 2.2.

This early work provided a most useful introduction, but did not take account of some additional effects which may be important. In particular it provided no frictional force in the case where the stream was incident obliquely.

One additional effect may result from the electron motion parallel to the boundary and perpendicular to **B** as in Figure 7.5a. In the case of the solar wind, the velocity of the electrons exceeds 10^9 cm/sec. This motion through the ambient plasma (if any) and through the nearly stationary protons may give rise to the *two-stream instability* (Section 3.4) and to the growth of electric space-charge clouds or waves of dimensions about 0.1 km and potential differences about 10^3 V (Piddington, 1960; Bernstein et al., 1964). The result is a general scattering or thermalizing of particles within the boundary layer so that instead of suffering specular reflection many remain trapped. For a stream incident at an angle to the normal, this effect clearly permits transfer of momentum parallel to the boundary, and so, a frictional interaction. In this way field lines may be moved parallel to the boundary (forced interchange motions), and perhaps into the geomagnetic tail.

When there are enough thermal electrons in the boundary layer with velocities comparable with the streaming velocity, the two-stream instability is suppressed. Meanwhile, however, the particle thermal energy has risen to about 10^3 eV or 10^7K, and even if the density does not exceed that in the solar wind, the pressure is about 10^{-8} dyne/cm^2, which equals the average magnetic pressure in the magnetopause. This inflation of the boundary layer by high-pressure plasma may cause "magnetic bubbles" to form and project across the boundary, an effect which would explain the results of Sonnerup and Cahill. Transport of these bubbles (still connected to the geomagnetic field lines) would explain tail growth.

As seen in Section 6.4, equilibrium is established between the geomagnetic field and the impinging plasma stream when the magnetic pressure $B^2/8\pi$ balances the stream pressure. The stability or otherwise of this equilibrium has been investigated by Fejer, Sen and by Lerche (1966, where earlier references are given). If these investigations are to be helpful in explaining the problems of tail growth and varying geomagnetic activity, then they must be related to the correlation studies of solar wind and geomagnetic activity discussed above (Section 6.5).

More recently Parker (1967) has discussed the possibility that the space-charge field developed by the more penetrating protons is neutralized, not by the accelerated electrons as in Figure 7.5a, but by electrons sucked up from the ionosphere. In this event the impinging electrons are bounced off the boundary layer as in Figure 7.5b. Suppose now that the impinging protons have a velocity component along the magnetic field lines, that is, into or out from the paper. They give rise to an entirely new current component within the boundary layer and parallel with the magnetic field. This current generates a new magnetic field component perpendicular to the original field, with a scale in the range $1-10^3$ km, and this layer of field cannot be contained and so is eroded away. One aspect of this boundary model concerns the effects in the ionosphere itself. A simple calculation, based on an ionospheric critical frequency of 5 MHz and average solar wind, shows that the electron requirement to neutralize the layer is about ten times the total electron content of the ionospheric strip connected to the boundary layer. Hence a 10% increase in solar wind flux will remove all ionospheric electrons. As we have seen in Section 3.5, electrons are not able to move across magnetic field lines anywhere in the ionosphere except as Hall drift. Hence the Pedersen current required to make up for electron flow upwards is provided by ions flowing outwards from the ionospheric strip which is emptied of both electrons and ions. Such an effect should be detectable by ionospheric measuring devices.

In discussing various models and theories of the geomagnetic boundary we must not overlook the possible importance of an interplanetary electric field, the existence of which has been stressed by Alfvén and discussed more recently by Rostoker and Fälthammar (1967). The interplanetary plasma moving with velocity $\mathbf{V}_i$ carries a magnetic field $\mathbf{B}_i$, and so an observer fixed in relationship to the geomagnetic cavity sees an interplanetary electric field $\mathbf{E}_m = -\mathbf{V}_i \times \mathbf{B}_i$.

It has been argued that the electric field is confined to the region of the solar wind and so can have no effect within the cavity, but in view of the apparent complexity of the wind-cavity interaction the possible effects of $\mathbf{E}$ may warrant further investigation, perhaps along the following lines.

Consider two systems of axes, one fixed in the magnetosphere and one fixed in the interplanetary plasma. In the first system one sees an electric field $\mathbf{E}_m$ which pervades interplanetary space, and in the second one sees an electric field $\mathbf{E}_i$ which pervades the magnetosphere. These fields are given by

$$\mathbf{E}_m = -\mathbf{V}_i \times \mathbf{B}_i \qquad \mathbf{E}_i = \mathbf{V}_i \times \mathbf{B}_m \qquad (7.1)$$

where $\mathbf{B}_m$ is the magnetospheric field. If $\mathbf{B}_i$ and $\mathbf{B}_m$ are opposite directed, then the electric fields are aligned with one another. This situation is shown in Figure 7.6 at the dawn boundary ($y = 0$) between the two

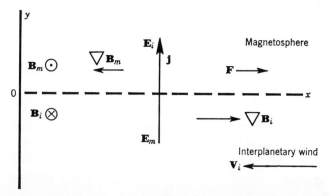

Fig. 7.6. The dashed line ($y = 0$) represents the boundary of the magnetosphere near the dawn meridian, viewed from above the north pole with the sun in the direction Ox. The magnetospheric field $\mathbf{B}_m$ is out from the paper and the interplanetary field $\mathbf{B}_i$ inwards. Electric fields $\mathbf{E}_m$, $\mathbf{E}_i$ are measured in systems of axes fixed in the magnetosphere and in the interplanetary plasma, respectively. A current $\mathbf{j}$ tends to flow inwards and give rise to other changes shown.

media, the magnetosphere being at the top and the sun in the x direction. Both electric fields are inwards, so that an ion moving across the boundary in either direction experiences an inward force; electrons are accelerated outwards. Thus random diffusion across the boundary tends to be converted to a current $\mathbf{j}$ flowing into the magnetosphere. The mechanical force $\mathbf{j} \times \mathbf{B}$ accelerates magnetospheric ions to solar-wind velocity and solar ions to magnetospheric velocity, and so constitutes a frictional force. Through Maxwell's curl of $\mathbf{B}$ equation 2.2, magnetic force gradients ∇B are developed as shown as part of the frictional interaction. The current **j** has ionospheric effects discussed in the following chapter.

In summing up the possible relationships between the various magnetopause effects discussed above and the observational data, the primary consideration seems to be the fact that almost all notable geomagnetic disturbance is associated with strong, southward magnetosheath fields. It seems that such a field is required to cause growth of the magnetotail, but unfortunately there now appear to be no less than four possible mechanisms.

(1) The mechanism of field annihilation and reconnection depends entirely on a southward magnetosheath field to provide tail growth.

(2) Inflation of the boundary layer by plasma which has been heated through the two-stream instability (Figure 7.5a) depends on a boundary layer current. Such a current must have maximum intensity when there is a southward external field, and may be zero for a northward field of strength equal to the internal field.

(3) The mechanism illustrated in Figure 7.6 depends on an external field.

(4) The interplanetary electric field E_m (Figure 7.6) may penetrate the tail neutral sheet and inhibit its decay, as discussed in the following section.

It does not seem possible at present to determine which of these effects is responsible for geomagnetic disturbance. Indeed, the complexity of such disturbance, described in the following chapter, is so great that there seems a possibility that two or more mechanisms may be important.

7.4   Motions of Magnetic Field Lines to and from the Tail

There is fairly conclusive experimental evidence of motions of magnetic field lines into the geomagnetic tail and back from the tail. In this section we discuss these motions further, in connection with the theoretical models which have been proposed and some observational data.

Motion to the Tail

Motion to the tail, although it may have several possible driving mechanisms, seems to be limited to two possible kinematic forms.

The form first suggested (Piddington, 1965; Atkinson, 1966) involves motion of unbroken field lines around one or both sides of the corotating magnetosphere. This *forced interchange* motion is illustrated in Figure 7.7, a particular tube of magnetic force being shown in

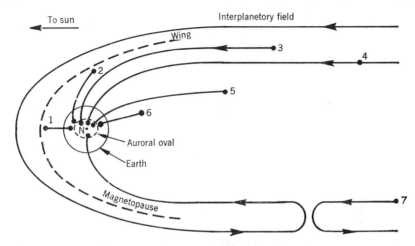

Fig. 7.7. A schematic representation of a geomagnetic field line or tube of force (viewed from above the north pole) as it moves through successive positions 1–4 into the tail and through positions 5 and 6 back to the corotating magnetosphere. Line 7 has connected with the interplanetary field after being drawn into the tail.

successive positions. Viewed from above the north pole a force tube occupies position 1, stretching from hemisphere to hemisphere, with the part of the tube lying in the line of sight shown as a dot. The successive positions of the intersection of the tube with the equatorial plane are shown by numbers 2–6 and the distortion of the line is also sketched. At position 4, after being extended far into the tail, the line reconnects at the dot and then moves through position 5 back to the corotating magnetosphere at 6. Each time the force tube moves, the region vacated is filled by another tube so that a flow pattern is set up. The corresponding flow in the ionosphere is also shown, being first into the auroral oval and then out again. Subsequently the force tube moves from position 6, through the magnetosphere to its initial position 1.

There is some evidence in favor of this pattern of flow in the poorly defined magnetopause at low latitudes near the dawn meridian: the "wing" of the magnetosphere. Gosling *et al.* (1967), measuring streaming protons, occasionally found a region some hundreds of kilometers thick where there appeared to be a mixture of streaming and non-streaming plasma. In the presence of a magnetic field, such a mixture indicates alternate filaments or sheets of plasma, some sharing the streaming motion of the magnetosheath. Magnetic field measurements by Heppner *et al.* (1968) show regions where the magnetosphere is so inflated that interchange phenomena are likely (see Section 4.3). These regions are near the dawn meridian and have thicknesses of several earth radii, and would seem capable of providing the required rate of flow to the tail. Again Fairfield and Ness (1967) found several cases where no magnetopause could be detected. Finally the observations of Speiser and Ness (1967) of a particularly strong field component across the tail neutral sheet and near the dawn side may indicate field lines moving into the tail, rather than returning.

A quite different flow pattern has been suggested by Dungey (1961, 1967) and is illustrated in Figure 7.8. An interplanetary field line moves

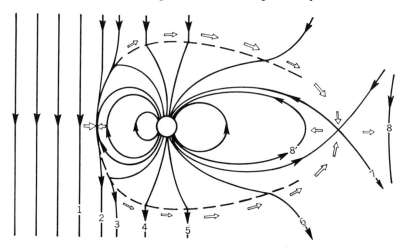

Fig. 7.8.   The formation of a geomagnetic tail by severing field lines at the magnetosphere boundary (on the solar side) and connecting them with interplanetary field lines. The numbers indicate the successive positions of a particular interplanetary line from a time prior to its connection with a geomagnetic line (1) to its release (8) after reconnection of the geomagnetic line.

from position 1 to position 2, where it breaks and connects with a geomagnetic line. It moves through positions 3–6, drawing the geomagnetic line into the tail. At position 7 it reconnects across the tail neutral sheet and the inner section moves back towards the earth. Some observational evidence in favor of this flow pattern may be provided by the results of Krimigis and Van Allen (1967), who find that solar protons invade the polar ionosphere very shortly after they arrive in the vicinity of the earth (but outside the magnetosphere). Figure 7.8 shows the polar ionosphere directly connected to interplanetary space by the magnetic field lines and so would seem to explain this rapid invasion. However, this figure is schematic and misleading, because a field line connected to the interplanetary field is stretched along the tail at the rate of about 300 km/sec or 170 R_E/hour (quiet conditions). As we shall see in Chapter 8, the foot of the field line probably takes several hours to cross the polar cap and so connection between the pole and interplanetary space is still through a tail a few hundred R_E long. Observations of polar cap particle invasion following a flare show rather complex patterns (Hakura, 1965) which do not seem to be easily explained in terms of either flow pattern.

Motion from the Tail

Motions of field lines from the tail appear to follow reconnection through the neutral sheet and in the manner already described. There is also the possibility that some lines which have been drawn into the tail in the pattern of Figure 7.7 are released by the solar wind and contract directly without reconnecting across the neutral sheet and losing their outer section.

Possible mechanisms of field annihilation and reconnection have been discussed in Section 4.2. The application of these mechanisms to the magnetotail have been discussed by Piddington (1965, 1967), Axford et al. (1965), Coppi et al. (1966), Speiser (1967) and Dessler (1968).

These discussions differ in the attention paid to particular aspects of the problem. Thus Speiser considers the detailed motions of particles under the influences of the main tail field, the neutral sheet field and an electric field from the dawn to the dusk side of the tail caused by motions of magnetic field lines into the neutral sheet. Particles are accelerated, turned towards the earth and ejected from the neutral sheet at small angles to the field lines. They have energies of some tens of keV and ejection occurs at distances of 50–500 R_E from the earth. Dessler's model, on the other hand, is based on the hypothesis that just behind the corotating magnetosphere the plasma density is so low that the field lines cannot be kept apart and so merging occurs, being complete within 30 R_E of the earth.

The question of whether reconnection occurs within 30 R_E or beyond 50 R_E or 500 R_E is important, because of energy considerations. As we have seen, energy in excess of 10^{24} erg is likely to be fed into the tail (as magnetic energy, of density $B^2/8\pi$) during a major storm. However, this energy is spread out to a distance of 1000 R_E and if the tail is "nipped off" at 30 R_E, then all the energy beyond 30 R_E is lost and only a few times 10^{22} erg is available. This is not enough to power the storm and aurora. The question seems to have been answered by

Mihalov *et al.* (1968), who find sporadic northward directed fields out to the maximum distance explored (80 R_E), and sporadic southward fields only beyond 30 R_E, and then chiefly near the side of the tail until distances beyond 60 R_E are attained. After reconnection, the northward field line moves inwards (Figure 7.1) and the southward line moves outwards so that some reconnection must occur beyond 80 R_E, and it would seem that most does occur beyond 60 R_E. From these results it would appear that the short tail model is not valid and that the energy available from conversion of magnetic energy may be adequate to power storms and auroras.

The energy available from the geomagnetic tail to power the Van Allen belt and aurora is proportional to the distance along the tail to the region of reconnection. Thus by delaying reconnection and increasing this distance the level of geomagnetic activity may be increased, even though the flow of field lines to the tail remains unchanged. This factor, which has not been generally appreciated, may be important in explaining the relationship between activity and the solar wind and field.

As we saw in the preceding section, an electric field E_m is seen in the solar wind by an observer fixed in relation to the earth. A southward interplanetary magnetic field provides a dawn-to-dusk electric field, and if this penetrates the tail neutral sheet, then it may inhibit reconnection. We might explain this effect in terms of the Hall drift B^{-2} ($E_m \times B$) of the tail plasma. This drift is into the neutral sheet, where accumulating plasma blocks magnetic merging. Another way of looking at the problem is that merging of lines across a neutral sheet is usually required to provide an electric field which drives the sheet current. When an electric field of external origin, E_m, is available merging is not necessary and so ceases.

As we saw in Section 4.2, magnetic merging may be accelerated or slowed by the introduction of hydromagnetic compression waves. A wave with field-strength gradient ∇B in the direction of the sheet current accelerates merging and one in the reverse direction slows merging. If these effects are significant in the magnetotail, then buffeting of the tail by turbulence or change in wind direction will be important.

References

Akasofu, S.-I., 1966, *Space Sci. Rev.* **6**, 21.
Anderson, K. A. and Ness, N. F., 1966, *J. Geophys. Res.* **71**, 3705.
Atkinson, G., 1966, *J. Geophys. Res.* **71**, 5157.

Axford, W. I., Petschek, H. E. and Siscoe, G. L., 1965, *J. Geophys. Res.* **70**, 1231.
Bame, S. J., Asbridge, J. R., Felthauser, H. E., Hones, E. W. and Strong, I. B., 1967, *J. Geophys. Res.* **72**, 113.
Behannon, K. W., 1968, *J. Geophys. Res.* **73**, 907.
Bernstein, W., Fredericks, R. W. and Scarf, F. L., 1964, *J. Geophys. Res.* **69**, 1201.
Cahill, L. J., 1966, *J. Geophys. Res.* **71**, 4505.
Carpenter, D. L., 1966, *J. Geophys. Res.* **71**, 693.
Coppi, B., Laval, G. and Pellat, R., 1966, *Phys. Rev. Letters* **16**, 1207.
Dessler, A. J., 1968, *J. Geophys. Res.* **73**, 209.
Dungey, J. W., 1961, *Phys. Rev. Letters* **6**, 47.
Dungey, J. W., 1967, *Imperial College (London) Res. Rep. SP*-67-4.
Fairfield, D. H., 1968, *J. Geophys. Res.* **73** (GSFC X612–68–124).
Fairfield, D. H. and Ness, N. F., 1967, *J. Geophys. Res.* **72**, 2379.
Feldstein, Y. I. and Starkov, G. V., 1967, *Planet. Space Sci.* **15**, 209.
Ferraro, V. C. A., 1952, *J. Geophys. Res.* **57**, 15.
Frank, L. A., 1968, *Earth's Particles and Fields*, B. M. McCormac, ed., Reinhold, New York.
Frank, L. A., Van Allen, J. A. and Craven, J. D., 1964, *J. Geophys. Res.* **69**, 3155.
Gosling, J. T., Asbridge, J. R., Bame, S. J. and Strong, I. B., 1967, *J. Geophys. Res.* **72**, 101.
Hakura, Y., 1965, *J. Radio Res. Lab. (Japan)* **12**, 231.
Heppner, J. P., Ness, N. F., Scearce, C. S. and Skillman, T. L., 1963, *J. Geophys. Res.* **68**, 1.
Heppner, J. P., Sugiura, M., Skillman, T. L., Ledley, B. G. and Campbell, M., 1968, *J. Geophys. Res.* **72**, 5417.
Krasovskii, V. I., Shklovskii, I. S., Gal'Perin, Yu. I., Svetlitska, E. M., Kushnir, Yu. M. and Bordovskii, G. A., 1961, *Artificial Earth Satellites* Vol. 6, Plenum Press, New York, p. 137.
Krimigis, S. M. and Van Allen, J. A., 1967, *J. Geophys. Res.* **72**, 4471.
Lerche, I., 1966 *J. Geophys. Res.* **71**, 2365.
McCormac, B. M., ed., 1966, *Radiation Trapped in the Earth's Magnetic Field*, D. Reidel, Dordrecht, Holland.
McDiarmid, I. B. and Burrows, J. R., 1965, *J. Geophys. Res.* **70**, 3031.
Meinel, A. B., 1950, *Phys. Rev.* **80**, 1096; also *Astrophys. J.* **113**, 50 (1951).
Mihalov, J. D., Colburn, D. S., Currie, R. G. and Sonett, C. B., 1968, *J. Geophys. Res.* **73**, 943.
Murayama, T. and Simpson, J. A., 1968, *J. Geophys. Res.* **73**, 891.
Ness, N. F., 1965, *J. Geophys. Res.* **70**, 2989.
Ness, N. F., Behannon, K. W., Scearce, C. S. and Caterano, S. C., 1967, *J. Geophys. Res.* **72**, 5769.
Ness, N. F., Scearce, C. S. and Caterano, S. C., 1967, *J. Geophys. Res.* **72**, 3769.
Ness, N. F., Scearce, C. S., Seek, J. B. and Wilcox, J. M., 1966, *Space Research* **VI**, 518.
O'Brien, B. J., 1963, *Space Sci. Rev.* **1**, 415.
Parker, E. N., 1967, *J. Geophys. Res.* **72**, 2315 and 4365.
Piddington, J. H., 1959, *Geophys. J. Roy. Astron. Soc.* **2**, 173.

Piddington, J. H., 1960. *J. Geophys. Res.* **65**, 93.
Piddington, J. H. 1965 *Planet. Space Sci.* **13**, 363.
Piddington, J. H., 1967, *J. Atmos. Terr. Phys.* **29**, 87; *Planet. Space Sci.* **15**, 733.
Rostoker, G. and Fälthammar, C. G., 1967, *J. Geophys. Res.* **72**, 5853.
Sonnerup, G. U. Ö. and Cahill, L. J., 1967, *J. Geophys. Res.* **72**, 171.
Speiser, T. W., 1967, *J. Geophys. Res.* **72**, 3919.
Speiser, T. W. and Ness, N. F., 1967, *J. Geophys. Res.* **72**, 131.
Van Allen, J. A., 1957, *Proc. Nat. Acad. Sci. (U.S.)* **43**, 57.
Van Allen, J. A., 1963, *Space Science*, D. P. Le Galley ed., John Wiley & Sons, New York.
Van Allen, J. A., Ludwig, G. H., Ray, E. C. and McIlwain, C. E., 1958, *I.G.Y. Satellite Rep. Ser.* No. 3, 73, National Academy of Science, Washington, D.C.
Vernov, S. N., Grigorov, N. L., Logachev, Yu. I. and Chudakov, A. E., 1958, *Dokl. Akad. Nauk. S.S.S.R.* **120**, 1231.
Zmuda, A. J., Heuring, F. T. and Martin, J. H., 1967, *J. Geophys. Res.* **72**, 1115.

CHAPTER 8

Geomagnetic Disturbance and Related Effects

This chapter deals with geomagnetic disturbances, or perturbations in the earth's magnetic field, and a number of related effects, notably auroras.

The study of the earth's magnetic field was made possible by the discovery of the directive properties of certain minerals, perhaps more than a thousand years ago. The development of these properties to give precision measurements was a result largely of their use in navigation, and led eventually to the discovery by Gilbert in 1600 that the earth itself is a giant magnet. Not unreasonably, Gilbert concluded that this magnetism was diffused throughout the planet and must remain constant, apart from geologic changes. However, only a few decades later Gellibrand found that this was not the case, and that the declination at London showed a slow *secular variation*. This variation was studied over the next three centuries and led eventually to an understanding of the origin of the magnetic field in internal fluid motions (Section 4.1), which themselves changed quite rapidly compared with geologic changes. A review of the changing magnetic field and its detection as remnant magnetization of rocks and baked earths (paleomagnetism) has been given by Nagata and Ozima (in Matsushita and Campbell, 1967).

Practical motives led mariners and surveyors to plot the geomagnetic field, and the results accumulated over the years were used by Gauss in 1839 to provide the first spherical harmonic analysis of the *surface field*. This provided a rigorous mathematical proof that the main field originates within the earth. Subsequent analyses revealed a decrease in magnetic moment of about 5% over the next century, and these and other investigations of the earth's field and its many fluctuations have been reviewed up to that period by Chapman and Bartels (1940).

The spherical harmonic analysis also determined the magnetic field distribution in the space around the earth, subject to possible modification caused by electric currents flowing in the atmosphere or beyond.

Neglecting these for the time being, the scalar potential of the field is represented by an expansion in spherical harmonics,

$$V = a \sum_{n=1}^{\infty} (a/r)^{n+1} \sum_{m=0}^{n} (g_n{}^m \cos m\phi + h_n{}^m \sin m\phi) P_n{}^m(\theta) \qquad (8.1)$$

where a is the radius of the earth, r, θ and ϕ are spherical polar coordinates and $P_n{}^m$ is the spherical harmonic of Schmidt (Chapman and Bartels, 1940). Knowing the various Gauss coefficients $g_n{}^m$ and $h_n{}^m$, one can calculate the field from the relationship $\mathbf{B} = \nabla V$. In general, measurements of $\mathbf{B}$ at many points on the earth's surface and in space are inserted in equation 8.1 and the coefficients are adjusted to give the best fit. Various sets of the coefficients have been derived to represent the field and different upper limits to the sum on n have been tried. For many years, a set of forty-eight coefficients derived by Jensen and Cain were commonly used in trapped radiation work. This set is being replaced by a representation based largely on the results from the satellite OGO-2.

Variations of the surface and space fields from the theoretical values are caused by ionospheric currents, currents generated by trapped particles and currents flowing on the boundary of the magnetosphere. When ionospheric currents are allowed for, the field is close to the theoretical, local-time independent value out to a few earth radii. Beyond that distance the warping of the geomagnetic field by the solar wind is a permanent, although varying, feature. A warped magnetospheric model has been proposed by Mead (1964) who gives values for twelve expansion coefficients. Reviews of geomagnetic field models have been given more recently by Vestine and by Price (both in Matsushita and Campbell, 1967).

The first term of equation 8.1 accounts for about 90% of the external field, and for many purposes it is sufficient to consider the field due entirely to this *centered dipole*. In this approximation the position and motion of a trapped particle may be specified very simply. First we specify a given field line by the geocentric distance r_0 where it cuts the equatorial plane. Then we specify the position of a point on that field line by the angle λ between the equatorial plane and the line from the earth's center to the point. We then have

$$\cos^2 \lambda = r/r_0 \qquad\qquad B = \frac{M}{r^3}\left(4 - \frac{3r}{r_0}\right)^{1/2} \qquad (8.2)$$

where r and B have been defined above and M is the dipole magnetic moment. It is useful to introduce a quantity L defined by $r_0 = L R_E$, where R_E is the earth's radius, and a second parameter Λ, which is the latitude where the field line in question cuts the earth's surface. We then have

$$\cos \Lambda = L^{-1/2} \qquad (8.3)$$

A particle trapped in this idealized field spirals along a field line and drifts around the earth at a fixed equatorial radius r_0, so that its motion is on a given L *shell* or at a given *invariant latitude* Λ. In real geomagnetic space, motion is more complicated, but as seen in Section 8.4 it is still possible to define L and Λ so that they remain approximately constant for particles of all energies (up to a given limit) and equatorial pitch angles.

In Section 8.1 we discuss *geomagnetic disturbances* of various forms which are caused by the solar wind; the disturbance morphology and current systems responsible are determined. A closely related phenomenon is the *aurora* which turns out to be a whole family of effects whose morphologies are described in Section 8.2. These are the most notable results of magnetospheric disturbance, but there are many minor by-products. These include transient particle events, electromagnetic and electric waves of all frequencies and numerous ionospheric effects, all discussed in Section 8.3.

In Sections 8.4 and 8.5 theoretical work on magnetospheric phenomena is discussed. This has been divided into the theory of individual particle motions and the theory of interaction of plasma and geomagnetic field. An extensive review of magnetospheric and related effects has been given by Shabansky (1968).

8.1 Geomagnetic Disturbance

Transient geomagnetic variations, which are far more rapid than the above-mentioned secular changes, were discovered in 1722 by Graham, who patiently recorded the tiny movements of a compass needle in London. On some days he noted slow and regular changes of declination, on others, irregular changes, sometimes much larger and faster. As other observatories were set up in different parts of the world, the *quiet-day* variations were mapped in detail and a quiet solar variation (Sq) was isolated. The immediate cause of Sq is an ionospheric current system which is driven by an electric "dynamo" field caused, in turn, by motion of the E-region plasma across the magnetic field lines. The gas motion results from uneven and varying heating of the atmosphere by solar radiation. The atmospheric dynamo theory is fairly well developed and has been discussed briefly in Section 3.5.

A second geomagnetic fluctuation present at all times is found to have a pattern fixed in lunar local time. This component (called L) is weaker than Sq by about an order of magnitude but is isolated by statistical methods. The L field is caused by a true (gravitational) atmospheric tidal motion induced by the moon. A review of both Sq and L is given by Matsushita (in Matsushita and Campbell, 1967).

When these two transient magnetic variations have been removed from the observational data, there remain irregular, and occasionally very much larger variations which are called the disturbance (D) field. Reviews of D-field morphology, differing to some extent in the data selected or stressed, have been given by Nagata (1963), Piddington (1964), Akasofu (1966) and Obayashi (1967).

The mathematical analysis of disturbance variation has led to the use of symbols having definite mathematical meanings. Thus Chapman has suggested a formal division

$$D(t,\ \theta) = D_{st}(t) + DS(t,\ \theta) \qquad (8.4)$$

where t is time measured from the onset of a storm (storm time), θ is longitude and all variables are also functions of latitude. D is the storm-time variation at an observatory, D_{st} is the axially symmetric part common to all observatories at the same latitude (that is, independent of local time), and DS is the remainder. This separation helps in describing the complex morphology of D and also provides inspiration in the location of the current systems responsible.

It is useful to plot the progress of a large disturbance or storm as two functions of storm time only, but to do so DS must somehow be represented in a form independent of latitude. Sugiura and Chapman (1960) have selected tables of DS for a particular latitude (30°) and converted them to a series of sine terms, having different phase angles. The first term is much larger than the others and so its coefficient may be used as a crude representation of world-wide DS. Figure 8.1 shows the resultant plot of D_{st} and DS averaged for a considerable number of great storms. The other quantities plotted are discussed below and the whole diagram allows us to follow the various storm effects from the onset at $t = 0$ until the storm subsides a few days later.

The D_{st} curve shows a sudden increase in the strength of the low-latitude field at $t = 0$. As we have seen, this is caused by increased pressure of the solar wind and its sea-level effect may be explained in either of two ways. The force of the solar wind is countered by the Lorentz force of an eastward electric current at the magnetopause and this, according to the Biot-Savart law, provides a northward magnetic field near the earth to reinforce the dipole field. From a

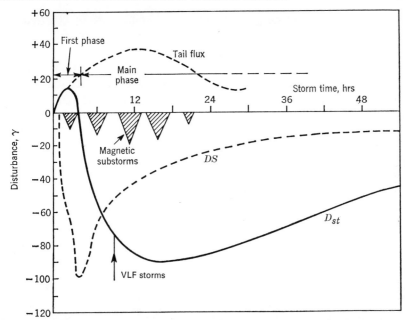

Fig. 8.1. Some geomagnetic storm phenomena plotted in storm time (from the onset of the storm). D_{st} is the longitude-independent change in the low-latitude field strength; a brief increase (first phase) is followed by a larger and more prolonged decrease (main phase). The DS disturbance field is a complex function of latitude and longitude, but has been represented by Chapman and Sugiura as a single quantity, as plotted. A series of magnetic substorms are shown schematically and on a reduced scale (this disturbance attains 1000 gammas in the auroral zone). The estimated rise and fall of the geomagnetic tail flux is also shown, plotted in arbitrary units.

magnetohydrodynamic point of view the field on the solar side of the earth is compressed between the wind and the earth, whose inertia is the ultimate obstruction to the wind (the magnetosphere freed from the earth would be blown away in a few minutes). The compression increases the field strength and so provides an increase in the average value around the earth.

It has long been thought that the large D_{st} decrease, called the main phase of the storm, must be due to a westward flowing current far from the earth and termed the *ring current*. The origin of this ring current was long a mystery until Singer (1957) pointed out that charged particles trapped within the magnetosphere would, because of their drifts and diamagnetic effect, provide such a current. With the discovery of Van Allen particles the theory had some observational basis and was developed quantitatively by Dessler and Parker (1959) and others and reviewed by Parker (1962). A secondary contribution to the D_{st} decrease will be made by the geomagnetic tail, by removing field lines from around the corotating magnetosphere and so allowing the field in that region to expand and weaken (Piddington, 1964). The criticism of this mechanism, that "the wind pressure must always dominate the outward tension," appears unjustified, because the outward tension, or frictional forces on the sides and tail of the magnetosphere, are more or less independent of wind pressure (Section 7.4). In addition, these forces act over the surface of the tail, along its full length of more than 1000 R_E, an area very much larger than that on which the wind exerts pressure.

The DS storm component grows to about the same amplitude as D_{st} (at higher latitudes it is much larger), but its peak is much sharper and it shows no reversal of sign. A most useful way of displaying world-wide DS is as an equivalent ionospheric current system, which is a current system whose magnetic effect is everywhere identical with the observed DS field. Although this current system is, initially, hypothetical, an examination of the field gradients reveals that most of it must be real current, because no more distant current system or systems could possibly give the rapid spatial fluctuations in DS. However, it is equally certain that all of the equivalent current system is not real current, because we know that currents in the magnetosphere and at the magnetopause are not symmetrical and so must contribute to DS. More important is the fact which has emerged from theoretical considerations, that there are real ionospheric current systems which make no DS contribution because their magnetic fields at sea level are cancelled by currents flowing up and down the magnetic field lines. Nevertheless, the equivalent current system is the basis of theories of the DS field and so must be carefully determined.

The most notable DS current system is that associated with an *auroral storm*, and termed DP1 (Obayashi, 1967), *negative bay* or *magnetic substorm*. A series of such substorms may occur during a great storm, and their waxing and waning are shown schematically, and on a greatly reduced scale, in Figure 8.1 by the hatched triangles. The ionospheric current system itself is shown in Figure 8.2, plotted on a projection of the northern hemisphere with latitudes and local times shown. The arrowed curves give the direction of current flow and their closeness of spacing the current density, typically 10^5 amp between lines. This plot is the classical DS system, derived statistically by observations at relatively few observatories, and it is fairly certain that the two "electrojets" (strong current concentrations) should be along the auroral oval described in Section 7.2. Individual substorms

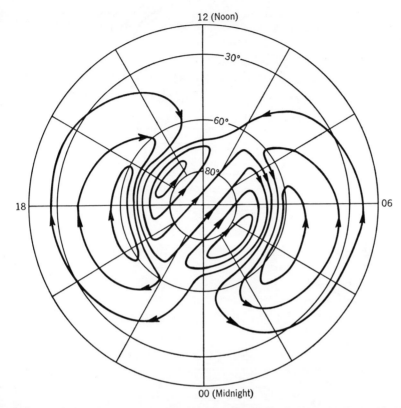

Fig. 8.2. An equivalent ionospheric current system plotted in (northern) latitude and local-time coordinates. If such a current system, with about 10^5 amp between adjacent lines, were to flow in the northern ionosphere it would cause the geomagnetic disturbance of the DP1 type (classical magnetic substorm).

differ greatly, and sometimes the current cells on the dusk side (the eastward electrojet) may be weak or absent. An example of such a current system is shown in Figure 8.3a (Fukushima, 1953). Yet another variation, due to Akasofu (1966) and his associates, is shown in Figure 8.3b. Here the westward electrojet flows around the auroral oval as far as mid-afternoon. The feature in common among these versions of the DP1 current system is a westward electrojet in the morning sector (from about midnight to some hours after dawn). This jet may carry 10^6 amp within a strip of width only a few hundred kilometers or less, the current density being about 10^{-2} emu/cm width of the ionosphere. This seems to be the main feature of DP1 requiring explanation.

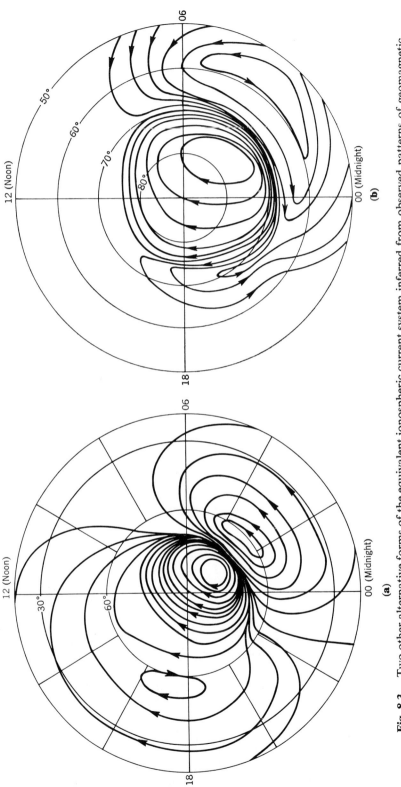

Fig. 8.3. Two other alternative forms of the equivalent ionospheric current system inferred from observed patterns of geomagnetic disturbance. (a) The twin current cell system found occasionally by Fukushima. The westward electrojet dominates this system, and the eastward jet with its pair of current cells near the dusk meridian (Figure 8.2) are weak or absent altogether. (b) Akasofu's current system, with a westward electrojet around the whole evening sector of the auroral oval.

A totally different ionospheric current system is found for conditions of *relative quiet*; outside storm time but with enough perturbation to be isolated from Sq. This is called DP2 and is shown in Figure 8.4 (see Nishida, 1968, who gives earlier references). The current flow is shown by the arrows **j** and those marked Vq are discussed below. Here the current flows in a dawn cell and a dusk cell which adjoin over the polar cap. It is interesting to note that the current flow is no longer concentrated along the auroral oval, but rather across it. The total current per cell is now only about 10^5 amp.

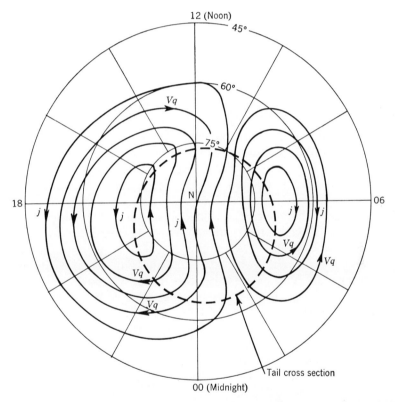

Fig. 8.4. The pattern of equivalent ionospheric currents during periods of slight to moderate magnetic disturbance appears to be quite different from that for storm conditions. This is the DP2 system, with 2×10^4 amp flowing between adjacent lines (Nagata and Kokubun, 1962). The arrows marked V_q denote a hypothetical convection pattern transmitted to the ionospheric electrons by moving magnetospheric field lines. The pattern is exactly the reverse of the current pattern and would account for the latter if all current were due to electron motions.

Any satisfactory theory of DS disturbance field must explain both DP1 and DP2 and indicate the reason for the drastic change. Theoretical work in this direction is discussed below, but a major difficulty is the uncertainty of the ionospheric current systems themselves. A requirement seems to be more observations at high latitudes, particularly near the auroral oval and further poleward.

8.2 Auroral Phenomena

It seems that the aurora has been studied even longer than the geomagnetic field, references to it having appeared in the literature and records of many nations over the past 2000 years. The name *aurora borealis* (northern dawn) was given in 1621, and it was about 150 years later that the first certain accounts of the similar southern *aurora australis* were given by Captain Cook. Later it was found that auroras appear simultaneously in both hemispheres and that they appear to be concentrated in auroral zones, or bands near 67° magnetic latitude. Meanwhile the close association between auroras and geomagnetic fluctuations had been observed by Celsius in 1741. The heights of auroras remained uncertain until early in the present century, when Störmer organized simultaneous photography from two stations connected by telephone. The heights vary from about one hundred to several hundreds of kilometers, which corresponds more or less with the ionosphere.

Other auroral characteristics which have been studied in great detail include their spectra, their morphology, correlations with other phenomena, and motions. Gradually it became certain that most auroral emission results from excitation and ionization of the atmosphere above 80 km, caused by fast particles moving down the magnetic field lines. These are mainly electrons with energies ranging from a few keV to some tens of keV, and they have been observed directly from rockets and satellites in and above the ionosphere. The total energy needed to sustain auroras around the world during average conditions is about 10^{18} erg/sec (O'Brien and Taylor, 1964). Some auroral emission is also provided by incoming protons, and this has a different morphology and spectral features. Reviews of auroral characteristics have been given by Störmer (1955), Chamberlain (1961) and Hultqvist (in Matsushita and Campbell, 1967).

As seen in Section 7.1, a radiation belt of energetic particles exists within the magnetosphere. The particle flux and spectrum shows a complex space–time pattern, and the outer boundary of this belt coincides more or less with the zone of occurrence of one type of aurora (Section 7.2). As seen above, auroral emission, like the flux of trapped particles, correlates closely with geomagnetic disturbance and so there is a clear connection between the two particle populations. At first it was thought that auroral particles were merely precipitating Van Allen particles, but it was soon evident that the energy requirement of 10^{18} erg/sec would deplete the belt within a few hours (O'Brien, 1964), so that this could not be the primary source. In fact, above a notable auroral display, that part of the radiation belt magnetically connected to the display would sometimes be emptied of trapped particles within a second or less (Winckler *et al.*, 1962). An even more conclusive argument against this origin of auroral particles is the fact that the flux of trapped particles increases when auroral precipitation takes place, so that a common source of the two populations is indicated. It is also found that auroral electrons are generally of lower energy than trapped electrons. Thus the common source of electrons must be capable of rapidly increasing the flux of trapped electrons of energy above 40 keV by a factor of 100, and the flux of auroral electrons (energy $\gtrsim$ 20 keV) by a factor of 1000, as observed.

In the present section the aurora is discussed as a manifestation of a magnetospheric disturbance, and as an aid in understanding this disturbance. It is clearly necessary, then, to distinguish between different types of auroras, which may have different origins. It is also necessary to consider the dynamic morphology of each type. That of discrete, visual auroras has been reviewed by Akasofu (1965) and mentioned in Section 7.2. These discrete auroras have been studied most intensively, partly because they are visually most notable and may be recorded by all-sky cameras. Of their various forms the multiple arc or multiple band are the most notable.

An example of a multiple-rayed band aurora is shown in Figure 8.5. Six bands are clearly shown, some of them having ray structure. The bands may extend for thousands of kilometers and yet have widths of 1 km or less. A similar auroral display would have occurred in the magnetic conjugate region in the southern hemisphere and it is interesting to speculate that the two arcs, band or multiple arcs or bands, represent a phenomenon similar to the pair of solar flare ribbons sketched in Figure 5.4 and shown less clearly in the photograph Figure 5.3.

During the past decade or so the study of auroras has vastly expanded from the previous visual and photographic records. They are now studied with photometers and other optical devices, by their radio wave reflections (radar auroras), absorption (riometers) and refraction (scintillation of signals from outer space), and above all by the direct observation of the particles responsible. They are also studied by means of secondary effects including the downward flux of x rays and radio waves. These results have changed our concepts of the aurora from a single phenomenon occurring at night usually near latitude 67°, to a whole family of phenomena occurring day and night at all latitudes down to about 40°, or lower during disturbance periods. The distributions of the principal auroral components have been given by Feldstein

Fig. 8.5. A photograph of a multiple-rayed band aurora taken at College, Alaska, by Dr. T. Ohtake (courtesy Geophysical Institute, University of Alaska, College, Alaska). The half-dozen bands resemble a set of parallel, illuminated curtains extending upward from a level of about 100 km.

(1963), Sandford (1964), Roach and Roach (1963), Eather (1967) and others, and reviewed by Akasofu (1965), Obayashi (1967) and Sandford 1966).

The morphology of various auroral precipitations may be unified to reveal four main types of auroras falling into four zones (Piddington, 1965). These zones are shown in Figure 8.6, which shows the northern ionosphere, viewed from above, in the same format of magnetic latitude local time as Figures 7.3 and 7.4. The auroral distributions are

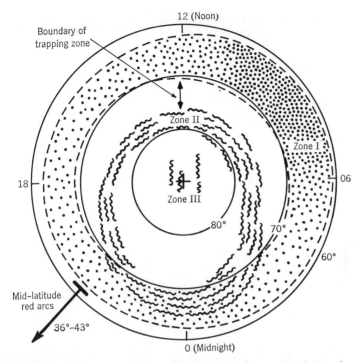

Fig. 8.6. A schematic representation of the pattern of auroral emissions for average conditions. It is plotted in the northern ionosphere in a magnetic latitude local-time coordinate system, and as in Figures 7.3 and 7.4 extends from the pole down to latitude 60°. Zone I (dotted) is the region of diffuse or mantle auroras, occurring mainly in the morning. Zone II (wriggly lines) is the auroral oval, or region of discrete visible auroras which predominate in the night sector. Zone III is the polar cap, and the zone of mid-latitude red arcs extends as shown.

schematic and do not stress diurnal variations. The zone of mid-latitude red arcs is indicated by the arrow, but it extends as a band right around the earth. A rather similar form of Zones I and II has been found by Hartz and Brice (1967). The characteristics of the precipitated particles, of the auroral emissions and associated phenomena are different for each zone and it is probable that the mechanisms responsible are also different for all four zones.

Zone I

Zone I defines the distribution of diffuse and subvisual "mantle" or "veil" auroras. It must be borne in mind that these are not the phenomena which have been observed for thousands of years and which are discussed above and in Chapter 7 (in connection with the magnetotail). Visually, mantle auroras are not very notable, but their properties are well documented by optical and other measurements. These properties are the following:

(1) The downcoming electrons have consistent, widespread, moderately intense fluxes and energies generally above 20 keV.

(2) The auroras are diffuse and steady.

(3) the auroras are associated with quasi-constant VLF radio emissions ("polar chorus") at frequencies below 2 kc/sec, and with continuous (pc) geomagnetic micropulsations (Section 8.3).

(4) the ionospheric absorption of radio waves is strong and slowly varying, and plasma irregularities (sporadic-E) develop in the range 80–90 km.

All of these phenomena have maximum intensity in the morning, and if we identify the geomagnetic tail boundary with Zone II of Figure 8.6 (the auroral oval of Section 7.2), then Zone I precipitation occurs well inside the corotating magnetosphere. The precipitation mechanism is not understood, but one suspects that it involves electric space-charge waves (Sections 3.4 and 4.4) which have a substantial velocity component along the magnetic field lines. Such waves will be generated at the front surface of the magnetosphere as shown in Section 7.3 and they will interact with newly trapped energetic electrons which drift eastward from their point of entry from the magnetotail. The precipitation thus tends to maximize in the morning. The difficulty in developing this, and other, electric field theories lies mainly in the difficulty of measuring electric fields in the magnetosphere (see Scarf in McCormac, 1968).

Zone II

Zone II of Figure 8.6 is the zone of discrete, visual auroras; it also defines Feldstein's (1963) auroral oval. The characteristics of its particles, emission and associated effects are listed as follows:

(1) The downcoming electrons arrive in intermittent bursts of intense fluxes of low-energy electrons ($E < 10$ keV), sometimes almost mono-energetic.

(2) The auroras are discrete, localized, bright and often rapidly fluctuating. Their forms are described as arcs, bands, rays, striations and so on, and arcs may extend east-west for thousands of kilometers,

with thicknesses less than 1 km. The rapid fluctuations have been referred to as "the merry dancers." Their heights are generally a little above 100 km.

(3) They are associated with bursts of relatively high-frequency (>4 kc/sec) VLF radio emissions ("auroral hiss"), and with impulsive (pi) geomagnetic micropulsations.

(4) Ionospheric absorption (measured by riometer observations of radio astronomy sources) is abrupt and highly variable, and plasma irregularities develop above 100 km.

(5) Unlike other auroras, those of Zone II are usually accompanied by "negative magnetic bays" (DP1), caused by a highly localized, mainly westward, electrojet flowing within the aurora itself. These phenomena generally show a maximum intensity in the night hemisphere and are very impulsive in character.

The auroral oval (Zone II) shown in Figure 8.6 corresponds to average conditions and as we have seen it shrinks and expands with waning and waxing magnetic activity. In the midnight sector it may move between latitudes above 70° and below 60°. During a great magnetic storm the only activity for a period may be a quiet arc below latitude 60°. Then within 5 min intense activity may spread from the original arc towards the pole, so that a series of twisted auroral arcs extend over a range of latitudes of more than 10° or a distance more than 1000 km. This is the auroral substorm, described and illustrated by Akasofu (1965). Some auroral arcs seem to surge westward while others break into fragments and move eastward. The phenomenon is accompanied by a magnetic substorm and other changes which last a few hours, after which the original situation is restored. The sequence of events may be repeated several times during a main-phase storm as shown by the hatched areas in Figure 8.1. Theories of discrete auroras are discussed in Sections 8.5 and 9.4.

Zone III

Zone III or polar cap auroras have different forms. Discrete auroras, similar to those of Zone II, have been observed with alignment towards the sun as shown in Figure 8.6, and with an inverse correlation with magnetic activity, in contrast to all other auroral types. Polar glow auroras follow a solar proton flare and are caused by the invasion of the polar cap ionosphere by protons and alpha particles in the energy range 1–100 MeV. These may cause a complete blackout of radio waves reflected from the ionosphere.

Mid-Latitude Red Arc

A relatively recently discovered (Barbier, 1958) auroral type is the mid-latitude red arc which occurs in two zones centered near magnetic latitude 53°, some 13° below the center of Zone I. These cover some hundreds of kilometers in both latitude and height and apparently occur in rings around the earth. Their emission is very stable and predominantly in the red lines of atomic oxygen (6300 Å, 6364 Å).

Red arcs lie at the feet of magnetic field lines with L values near 3 (equation 8.3), which corresponds with the center of the storm-time radiation zone. This, together with the relatively modest energy requirements, suggests that they may be energized in some way by the belt. Mechanisms of energy transfer which have been suggested include thermal conduction, direct precipitation of energetic particles, and electric fields (for references, see Cole, 1965). A major difficulty of these theories may be the narrowness of the arcs, about 300 km between half power points, which projects into the magnetosphere to about 2000 km in the equatorial plane. The radiation belt seems to be much wider than this, and so an explanation is required of the localization of the energy transfer mechanism itself.

8.3   Other Particle-Field Phenomena

In this section we review briefly a number of phenomena observed in the magnetosphere and ionosphere which may be regarded as byproducts of the main disturbance. These are transient particle and field events, and ionospheric disturbances additional to the auroral events discussed above. As we shall see, these effects represent relatively small energy transfer and are not important in the energetics of the storm. Nevertheless, they provide fascinating studies in their own rights and also help in understanding the storm mechanism.

Transient Particle Events

Transient particle events cannot be isolated completely from the main auroral and radiation belt changes in flux. However, whereas the characteristic times (rise time or decay time) of the main events may range from a few hours down to a minute or less, those of the transient events are minutes, seconds or even small fractions of a second. Transient events also tend to be more localized in space, although they occur throughout the magnetotail, the corotating magnetosphere and particularly in the transition, auroral region. Rapid changes in particle flux or spectrum may be observed by counters in rockets or spacecraft, or they may be studied indirectly at balloon levels by the x radiation which they generate in the lower ionosphere (Brown, 1966). Most of the measurements have been concerned with electrons of energies greater than 20 keV, because these are easier to study than lower-energy electrons or protons.

The slowest events which we might classify as transient are variations in the flux or spectrum of trapped electrons (energy > 20 keV) within periods of 1–20 min. These are associated with hydromagnetic waves propagating through the magnetosphere. Other slow particle transients are those associated with the bow shock wave which lies beyond the magnetopause. These "bow shock spikes" are seen as sudden increases of electron flux by as much as three or more orders of magnitude, lasting 10 min or more. They are associated with in-and-out motions of the magnetopause caused by fluctuating solar-wind pressure. Other "spike" events, this time highly localized in a plot in latitude and longitude, are shown in Figure 7.4. These and other magnetotail transients are discussed in the following chapter, where theories of their origin are reviewed.

Next in order of rapidity are fast pulsations in the fluxes of trapped and precipitated particles, having periods 5–10 sec in the early morning hours (0200 to 1000 local time) and 15–40 sec around mid-day. These are associated with micropulsations (discussed below) of the same periods. Yet faster phenomena are microbursts of precipitated, and less frequently trapped, particles with duration about 0.25 sec, spaced about 1 sec apart in groups. These are also associated with micropulsations, of frequency 5–100 c/sec. The fastest observed particle effects are fast bursts (there is not general agreement about the terminology and these are sometimes called microbursts) having characteristic times of 0.005–0.025 sec. These precipitation events occur at a rate of 2/sec at night and 0.02 /sec in daytime.

Observational results have been discussed in more detail by Parks *et al.* (1968) and wave-particle associations and references are given by Anderson (in McCormac, 1968).

Electromagnetic Fields

The spectrum of electromagnetic fields in and below the magnetosphere extends in frequency far above the part which is called geomagnetic disturbance. Starting with frequencies around 10^{-3} c/sec, these fields are hydromagnetic waves or oscillations of different types called micropulsations, which range in frequency up to about 5 c/sec. They occur as irregular fluctuations in the field vectors (pi) or as mainly continuous patterns (pc). They are further divided according to frequency or period, so that pc1 includes periods 0.2–5 sec, pc2,3 covers the range 5–45 sec, and so on. As we have seen in Section 8.2, pc oscillations tend to occur in auroral Zone I (steady, diffuse auroras), while pi oscillations are a feature of Zone II (discrete auroras). A detailed discussion of micropulsations has been given by Campbell (in Matsushita and Campbell, 1967).

Overlapping the micropulsation spectrum are ELF radio waves with frequencies 3–300 c/sec, and then VLF radio waves with frequencies up to 30 kc/sec. Again, these radio emissions show a complex distribution of intensity and spectrum. However, their main features may be summarized by dividing them into two spectral types which also have different distributions in the latitude local-time plot.

The first radio emission type is called "chorus," and consists of closely spaced randomly occurring discrete noise bursts, usually rising in frequency in the audio range 0.5 kc/sec upwards. As we have seen in Section 8.2, this type is associated with Zone I auroral precipitation. The second radio emission type is called "hiss," which has a broad spectrum and when observed in the auroral zone may extend from a few kilocycles per second to the VLF limit. "Auroral hiss" is associated with Zone II auroral precipitation.

It is well known from theory that electrons and ions spiralling along the geomagnetic field lines will couple closely to electromagnetic waves. For example, the anisotropic Alfvén (hydromagnetic) wave with left-hand circular polarization and frequencies near the ion gyro frequency couples powerfully with spiralling ions. Similarly, the so-called radio "whistler" couples with spiralling electrons. As the particle moves along the field line its gyro frequency changes over a range of perhaps 500 to 1 and so there is a good chance of coupling somewhere (see Kennel and Petschek, 1966). The close association observed between particles and waves confirms the theoretical inference that either the particle motions create the waves or the waves cause the particle motions. The problem is which causes which.

A fairly decisive answer seems to be provided by comparing the power spectra of the fields and particles as in Figure 8.7. Here the spectrum is derived from the average power flux during a precipitation-noise "event." An event, which may be a particle microburst with associated radio or hydromagnetic emissions, has two characteristic periods or frequencies. These are the period occupied by the whole event and the wave period; the black dots correspond to the latter. Thus event A comprised a VLF burst of 8×10^{-7} erg/cm^2/sec in the range 0.5–7 kc/sec (Gurnett and O'Brien, 1964). The coincident electron precipitation event and aurora is shown at A', with a period

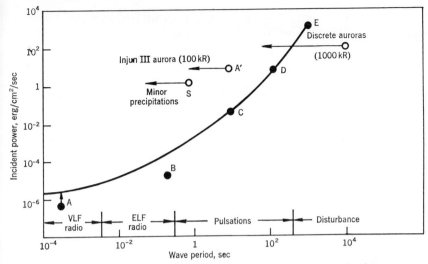

Fig. 8.7. A plot of the power spectrum of electromagnetic waves in the magnetosphere for wave periods 10^{-4} sec to a few hours. Electromagnetic "events" A to E, ranging from radio bursts to magnetic substorms, are described in the text. The open circles represent corresponding particle precipitation events. Apart from discrete auroras, precipitation events generally represent much larger power dissipation than the associated wave events and so it is argued that the wave events are not responsible for the precipitation events.

corresponding to the envelope of the event. The arrow above A shows the observed range of VLF power flux. Event B is typical of fast hydromagnetic pulsations and events C and D of increasingly long-period pulsations. Finally, event E is a notable magnetic "bay" disturbance which may be regarded as an ionospheric current of 10^6 amp driven by an electro-magnetic force of 10^4 V. This electrojet has an extent of about 10^3 km by 10 km, so that the power flux density is about 1000 erg/cm^2/sec. The electromagnetic power spectrum was drawn as shown on the basis of these data, and on subsequent testing is found to give a reasonable fit to other data.

The wave periods allotted to precipitation events (open circles) correspond to their durations or modulation envelopes, and so these events need not be compared with that part of the spectrum curve of the same wave period. The power flux density ranges from minor events marked S (1 erg/cm^2/sec) to very bright discrete auroras (>100 erg/cm^2/sec). It would seem then, that all electromagnetic events with wave periods of less than about a minute are caused by much more powerful particle events, rather than the reverse. However, even if we accept this interpretation, the role of radio waves in the magnetosphere may be important. They may receive energy from one group of particles and pass it to another without storing more than a tiny fraction of the total energy.

Electric Space-Charge Waves

Electric space-charge waves and oscillations may be another important ingredient in magnetospheric dynamics. Such waves may efficiently accelerate particles along the field lines and they may also couple with electromagnetic waves because of irregularities in the ambient plasma. They provide a further factor for consideration in relation to the above particle-field problem, and so further complicate the theory. They also present an additional difficulty in their observation; electromagnetic waves may be measured very easily by a magnetic loop but electrostatic waves require a probe which is affected by the presence of ambient plasma and the spacecraft itself.

In a plasma, electric fields of the form

$$\mathbf{E} = -\nabla\phi \qquad \phi = \phi_o\cos(\mathbf{k}.\mathbf{r} - \omega t) \qquad (8.5)$$

may occur and must satisfy Poisson's equation. At zero temperature there is only one, non-propagating solution, but in a hot plasma an electron wave and an ion wave may propagate (Section 3.4). Propagating along the magnetic field, these waves have no magnetic vectors and may only be detected by instruments which measure the electric components of waves. The phase velocities of these waves are important in the acceleration of particles trapped in the electric potential trough of the wave. Electron waves have phase velocities up to and exceeding the velocity of light and so may be very effective in this way. On the other hand, ion waves (which have been detected) have phase velocities given approximately by

$$\frac{\omega}{k} = \left(\frac{\kappa T_e}{M}\right)^{1/2} \qquad (8.6)$$

where κ is Boltzmann's constant, T_e is the electron temperature and M is the ion mass. For $T_e = 10^3 K$ and when ions are all protons, the velocity is only about 3 km/sec.

Reviews of electric wave measurements in the magnetosphere have been given by Scarf and by Storey (both in McCormac, 1968).

Ionospheric Disturbances

Ionospheric disturbances associated with geomagnetic disturbance have numerous forms and range in extent from world-wide density changes, to some forms of auroral sporadic E with dimensions of order

1 km. At the time of a solar flare there is a sudden ionospheric disturbance (SID), or increase in ionization density over the sunlit hemisphere caused by a sharp increase in solar x radiation. Later there may
be polar cap absorption (PCA), which is one form of Zone III auroral
events, and is caused by the invasion of the polar caps by protons with
MeV energies. These are the only ionospheric disturbances which are
reasonably well understood, although there are fairly satisfactory
theories of some other types of disturbance. The disturbance morphology and theories have been reviewed by Reid (in Matsushita and
Campbell, 1967).

The time variation of the maximum electron density in F region
(the *ionospheric storm*) may be divided into two components, one
longitude-independent (D_{st}) and one a function of local time (DS)—
just as for the magnetic disturbance components of Section 8.1. The
D_{st} component at high and middle latitudes shows an initial small
increase followed by a large decrease; near the equator there is an
increase followed by a small decrease. The DS component is mainly
diurnal in character and also has a minimum at mid latitudes with
larger values at high latitudes and near the equator. It is fairly clear
from these studies that F-region storms involve no great increase in
the total ionization content of the whole atmosphere, but rather a
redistribution caused by differential drift motions. The main drift
motions seem to be Hall drift caused by electric fields transferred down
from the magnetosphere, where they are an important part of the geomagnetic disturbance field. Additional motions of the plasma along
the magnetic field lines is caused by pressure gradients and gravitational
force. The overal picture is not yet clear.

In addition to the large-scale ionospheric storm, numerous smaller-
scale electron and ion density irregularities occur in the ionosphere.
The origin of these irregularities lies in general in differential electrodynamic drift and so is intimately related to their general drift motions,
discussed in Section 3.5.

8.4 Theory of Individual Trapped Particles

As seen from equation 2.1, the motion $\mathbf{v}$ of a particle in a magnetic
field $\mathbf{B}$ is given by

$$\dot{\mathbf{v}} = \frac{e}{m}\, \mathbf{v} \times \mathbf{B} \qquad (8.7)$$

the acceleration being at right angles to the particle velocity and to
the field. In a uniform field the trajectory is a helix whose axis is

parallel to the field line, and the motion may be regarded as gyration about a guiding center together with a translation of the guiding center along the field line. If the field is nearly uniform over distances equal to the radius of the helix, the motion will be approximately helical; however, the perturbations resulting from the field gradient are extremely important and are entirely responsible for geomagnetic (and other) trapping.

Three Motions

Trapped particles may be considered to have three separate motions. As the particle approaches a converging or mirroring field, the force acting at right angles to the field develops a component parallel to the guiding center trajectory and if the field converges sufficiently, this leads to reflection, as explained in Section 4.4. Such reflection takes place in the geomagnetic field since the field strength increases with increasing latitude as the particle moves along a field line. Thus the particle bounces from hemisphere to hemisphere as it spirals along a field line.

Other important perturbations in the helical motion result from the decrease in B with distance from the earth's center, and the centrifugal force experienced by the particle as its guiding center follows the curvature of the field line. Both these factors cause the radius of gyration to be slightly larger during the outermost half of the gyration, so that on each cycle the particle is displaced slightly in longitude, electrons moving eastward and positive ions westward. The equation for the drift motion $u_\perp$ across the magnetic field lines is

$$\mathbf{u}_\perp = -\frac{\mathbf{B}}{B^2} \times \left\{ \mathbf{E} - \frac{\mu}{e}\left(1 + \frac{2v_\parallel^2}{v_\perp^2}\right)\nabla B \right\} \qquad (8.8)$$

where $\mu = mv_\perp^2/2B$ is the magnetic moment of the particle, $v_\parallel$ and $v_\perp$ are its velocity components along and perpendicular to the field, and a perturbation electric field $\mathbf{E}$ has been introduced for completeness.

Typical periods for gyration (T_g), bounce (T_b) and drift around the earth (T_d) are given for particles confined fairly close to the equator at a geocentric distance 2 R_E.

	$T_g sec$	$T_b sec$	$T_d min$
proton (10 MeV)	0.02	0.42	1.7
electron (1.0 MeV)	2×10^{-5}	0.13	40

In each case there are large differences between the three periods and this makes it possible to consider the overall trajectory as three separate motions.

In general it is not practical to integrate equation 8.7 and the quantitative theory of trapped particles is based on the adiabatic invariants derived by Northrop and Teller (see Northrop, 1963). These are: 1st invariant = magnetic moment $\mu = mv_\perp^2/2B$ and 2nd invariant = "longitudinal" or "integral" invariant $J = \int mv_\parallel ds$, the integration being over a complete north-south bounce. Strictly speaking these are not invariants, but rather the first terms of asymptotic series which are the true adiabatic invariants. However, if the particle gyro radius is very small compared to the dimensions of significant field perturbations, if no electric field is present, and if the magnetic field is static, then terms beyond the first are negligible. The constancy of μ governs the pitch angle θ as the particle moves along a field line. Since $v_\perp = v \sin \theta$ and v is constant, we have $\sin^2\theta/B =$ constant, and knowing θ and B at any point the value of B at the mirror point ($\sin \theta = 1$) may be determined. The integral invariant governs the magnetic field lines to which the particle drifts on its way around the earth (whose field is not a dipole field). At a given longitude, the particle will select the field line such that J retains its earlier value.

If the magnetic field varies so slowly that there is only a small change during a drift period T_d, then a third invariant is a useful concept: 3rd invariant = flux Φ = magnetic flux enclosed by drift. Thus if the overall geomagnetic field were to increase slowly, the shell traversed by a particle would contract towards the earth.

As we shall see, each of these "invariants" may change in the presence of suitable electric fields or time-dependent magnetic fields. The results are discussed below, and meanwhile we consider the significance of non-dipole terms in the earth's field.

Non-Dipole Terms

Non-dipole terms in the earth's field cause the trapped particle (radiation) belts to lose their azimuthal symmetry. One effect results from the magnetic anomalies where the surface value of the field is low. From the mirror equation, particles are reflected at the same value of B and in an anomaly their mirror points descend in altitude. This effect is observed above the South American anomaly, where the radiation belt approaches closest to earth.

An effect of azimuthal asymmetry which may be important is called shell-splitting. In a dipole field, all particles which initially mirror on a given field line trace out the same magnetic shell irrespective of their mirror points. In an asymmetric field, particles mirroring at different altitudes trace out different shells, although they all return to their original field line.

The earth's asymmetric field has provided a great deal of labour in comparing particle fluxes at different longitudes, and in the development of suitable coordinate systems. The most successful scheme is that of McIlwain (1961), which is an attempt to specify the magnetic shells on which particles move. The difficulty, of course, is that particles of different pitch angles may be on a particular shell at one longitude but move to slightly different shells as they drift (shell splitting). McIlwain's coordinate system, which we will not describe here, minimizes this error and provides a parameter L such that $L = 3$ refers to a magnetic shell whose equatorial crossing is about 3 R_E from the earth's dipole. The second coordinate used in conjunction with L is the field intensity B, and L may be replaced by the (approximately) invariant latitude Λ according to equation 8.3.

Acceleration and Diffusion

The most important theoretical aspects of trapped particles concerns their accelerations and diffusion across L shells. There are various mechanisms available, but all depend on violation of one or more of the adiabatic invariants to provide either acceleration or diffusion. They are conveniently summarized in Table 8.1 (following Roederer).

Table 8.1   Acceleration–Diffusion Mechanisms

Interacting effect	Conserves	Violates	Changes	Causes
Resonant Cyclotron acceleration by VLF or ELF wave	field line	μ and J	θ or $v_{\parallel}/v_{\perp}$	pitch angle diffusion
Bounce acceleration by micropulsations or electric space-charge	μ and field line	J	v	pitch angle and radial diffusion
Drift acceleration by recurrent ionospheric or magnetospheric electric fields	μ and J	Φ	v	radial diffusion
Stochastic Coulomb interactions with the atmosphere	field line	μ and J	θ or $v_{\parallel}/v_{\perp}$	pitch angle diffusion
Compression of the magnetosphere	μ and J	Φ	v	radial diffusion

A general discussion of particle acceleration has been given in Section 4.4, and the nature of the electromagnetic (and electric) fields available for acceleration in the magnetosphere has been given in Section 8.3. We will discuss briefly the interactions listed in Table 8.1.

A particle spiralling along a field line may see an alternating electric field due to an electromagnetic wave. If the Doppler-shifted wave frequency matches the cyclotron frequency, then $v_\perp$ may be steadily increased or decreased. The likely waves are VLF radio waves such as whistlers, or the lower-frequency ELF radio waves. Change of $v_\perp$ changes the pitch angle and may accelerate or retard article precipitation.

We have seen how first-order Fermi acceleration occurs when a particle is bouncing between two approaching mirrors. The same type of acceleration occurs when one mirror oscillates in synchronism with a bouncing particle, so that it is always approaching the second mirror as it reflects the particle. Another way in which $v_\parallel$ may be increased is by electric space-charge waves moving along the magnetic field line. It is possible that such acceleration does not require resonance but is effective over a relatively short distance, say 10^9 cm. Longitudinal electric fields of about 10^{-5} V/cm have been observed and if these travelled (as part of a wave) at the appropriate speed they could impart energy of about 10^4 eV.

Finally, we recall that there are a number of large-scale electric fields which appear in the magnetosphere, some due to space-charge accumulations and some to changing magnetic configurations including interchange motions and compression of the magnetosphere. All of these are capable of accelerating or decelerating particles, depending on their trajectories. Acceleration may be considered as resulting from drifts to regions of lower potential or as betatron acceleration.

A great deal of detailed work has been done on the subject of particle diffusion, which of course is closely related to that of acceleration. Discussions of various aspects of this problem have been given by Roederer, Fälthammar, Haerendel and by Roberts (all in McCormac, 1967).

8.5 Early Theory of Disturbance and Aurora

Modern theory of geomagnetic disturbance and aurora gradually developed from the early particle theory of Störmer, the electric field theory of Birkeland (1908) and Alfvén's terrella experiments (see Block, 1966, for references), and primarily the solar corpuscular stream theory of Chapman and Ferraro. Interpretation of the observational disturbance data was made possible by the method of harmonic analysis and greatly aided by the introduction of equivalent current systems located in the ionosphere. All of this work has been referred to above and discussed in great detail by Chapman and Bartels (1940).

Reviews of more recent work have been given by Parker (1962). who deals with magnetic effects of the radiation belt, Nagata (1963), Piddington (1964), Akasofu (1966), Obayashi (1967) and a number of authors in Matsushita and Campbell (1967). Here we will review briefly a number of ideas contained in these theories, which retain their value even though the theory as a whole often required revision.

(1) Singer's idea that a *radiation belt* of hot plasma would inflate the magnetosphere and so account for the long-sought, hypothetical ring current has been developed theoretically and confirmed by observational results.

(2) Second, there was the concept that *atmospheric winds* would move plasma across the field lines and so provide electric fields and currents. This idea was applied successfully to the quiet-time tidal motions (the dynamo theory), but failed when applied to the disturbance current system. Nevertheless, it drew attention to the requirement of a source of great power (10^{18} erg/sec) which must lie, presumably, above the ionosphere. It also stressed the relationship between the currents and plasma drift motions and led to the use of observed auroral drift motions as a guide to the motions of electrons and of magnetic field lines (for a review, see Cole, 1966).

(3) Third, there was the concept of disturbances originating at the boundary of the geomagnetic cavity and being transmitted to earth by hydromagnetic action (Dungey, 1955; Kato and Watanabe, 1955; Lehnert, 1956; Dessler, 1958; Piddington, 1959).

(4) Fourth, there was the concept of *frictional interaction* between the solar wind and the geomagnetic cavity. This might form a long magnetotail as described in Chapter 7, or it might bend the magnetic field lines from their meridian planes and send magnetic twists down to the ionosphere (Piddington, 1960) as a form of forced interchange motion (Section 4.3). As we have seen, a frictional interaction of any form implies the existence of an electric current entering the magnetosphere on the dawn side and leaving on the dusk side. This current must flow down the field lines and so arrive in the ionosphere near the auroral oval; it traverses the ionosphere and leaves along another tube of force on the dusk side. Positive space charge accumulates at the foot of the dawn force tube, and negative space charge near the foot of the dusk tube. This potential field causes Hall current to circulate in two cells which have the same form as the equivalent current system DP2

shown in Figure 8.4. If the similarity is based on true identity, then we must account for the fact that the ionospheric Pedersen current is not revealed by sea-level magnetic perturbations. The reason is that the Pedersen current is part of a hydromagnetic disturbance within the magnetosphere, mainly in the form of a twist like that of Figure 3.2. This cannot propagate in the non-conducting lower atmosphere, whereas the Hall current cells generate magnetic field perturbations which may propagate in vacuo.

(5) Fifth, there was the concept of large-scale magnetospheric *convection* (Axford and Hines, 1961) resulting from magnetic interchange motions (Section 4.3) driven by frictional interaction with the solar wind. In this work, deformation of the force tubes was neglected, and instead an ad hoc flow pattern consisting of two closed convection cells was set up. This was a somewhat different approach to the use of a frictional force and gave a more realistic convection pattern, but neglected the flow of current and power down the magnetic field lines, to power the ionospheric current system. Large-scale magnetospheric convection has been discussed further by Hines (1964).

(6) Sixth, there was the concept of the *electric polarization* of a radiation belt, as a result of its asymmetry. Suppose, for example, that the radiation belt extended only around the night side of the earth, as it might if the fast particles forming the belt were introduced from the

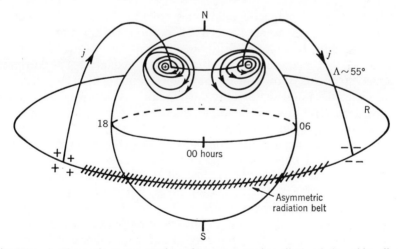

Fig. 8.8. A schematic representation of an asymmetric radiation belt and its effects. The earth is shown with local times marked, and a ring R at a few earth radii. The belt is shown hatched, the resultant space charge by plus and minus signs, and magnetospheric currents by arrows marked *j*. Ionospheric space charge develops and drives Hall currents in two cells as shown.

tail. This situation is shown schematically in Figure 8.8. An imaginary ring R around the earth is filled with energetic particles in the night sector (hatched length), and as a result of westward drift of protons and eastward drift of electrons, space charge develops as shown. This causes current j to flow down and up magnetic field lines and so polarizes the ionosphere in a manner somewhat similar to that described above. As before Hall current flows in cells as shown, but the currents are reversed and the cells are at latitudes well below the auroral oval. These effects of an asymmetric radiation belt were pointed out by Fejer (1961) and have been discussed in some detail by Cummings (1966) and others. The asymmetric belt is thought to be an important part of the magnetospheric storm and is discussed in this connection in the following chapter.

Electric polarization may also develop as a result of magnetospheric convection or interchange motions, together with irregularities in ionospheric ionization density N (Piddington, 1964). If the feet of the field lines move with velocity V, then they transport the electrons with almost the same velocity (in the lower E region) so that

$$\nabla.(N\mathbf{V}) = N\nabla.\mathbf{V} + \mathbf{V}\nabla N \qquad (8.9)$$

The result is a divergence of electron density, and since the ions cannot follow the field lines (because of collisions), electric space-charge fields develop.

(7) Motion of magnetic field lines to and from the magnetotail must also involve some pattern of motion within the corotating magnetosphere. In Figure 8.9 a plausible pattern of flow is shown by the dashed lines marked V_q (lines V_s are mentioned later). The corotating magnetosphere is shown in equatorial section as a circle marked in local time; the night sector boundary is not well known and is shown as dot-dash. The motions V_q represent those of points where the field lines intersect the equatorial plane; essentially the flow is from rear to front to compensate for the external motions (not shown) from front into the tail and tail back to the rear.

We may now visualize the motion of the foot of a field line in the northern hemisphere, on the assumptions that the line is always part of the configuration shown in Figure 7.1, and that the motions in the equatorial plane are those above. The resulting pattern is that shown as V_q in Figure 8.4 and will be followed by the ionospheric electrons. We know that ionospheric Hall current is carried by electrons alone, being attached to the magnetic field lines while the ions are held nearly stationary by collisions. Hence the DP2, or quiet-time, ionospheric equivalent current system is explained in terms of the V_q drift pattern,

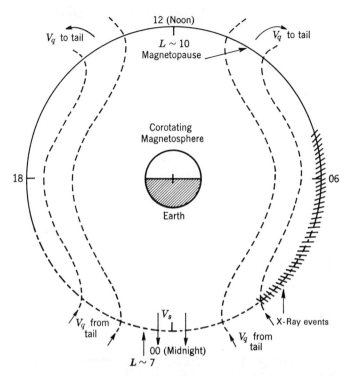

Fig. 8.9. The convection pattern V_q of plasma and field lines in the equatorial plane of the corotating magnetosphere (shown as a circle marked in local time). This section of the flow pattern is complementary to that shown in Figure 7.7 of motion from the front of the magnetosphere into the tail and back to the rear. Motion V_s corresponds to a substorm (Section 9.2).

provided the effects of Pedersen currents may be neglected.

The convection pattern V_q in the corotating magnetosphere (Figure 8.9) and its counterpart, projected downwards along the magnetic field lines to the ionosphere (Figure 8.4) is taken as the basic *steady flow pattern* at times of small, but not storm, disturbance.

References

Akasofu, S.-I., 1965, *Space Sci. Rev.* **4**, 498.
Akasofu, S.-I., 1966, *Space Sci. Rev.* **6**, 21.
Alfvén, H., 1955, *Tellus* **7**, 50.
Axford, W. I. and Hines, C. O., 1961, *Canad. J. Phys.* **39**, 1433.
Barbier, D., 1958, *Ann. Geophys.* **14**, 334.

Birkeland, K., 1908, *Norwegian Aurora Polaris Expedition* 1902–1903, Christiana, H. Aschehong 1908–1913, Vols. I and II.

Block, L. P., 1966, *J. Geophys. Res.* **71**, 858.

Brown, R. R., 1966, *Space Sci. Rev.* **5**, 311.

Chamberlain, J. W., 1961, *Physics of the Aurora and Airglow,* Academic Press, New York.

Chapman, S. and Bartels, J., 1940, *Geomagnetism,* Clarendon Press, Oxford.

Cole, K. D., 1965, *J. Geophys. Res.* **70**, 1689.

Cole, K. D., 1966, *Space Sci. Rev.* **5**, 699.

Cullington, A. L., 1958, *Nature (London)* **182**, 1365.

Cummings, W. D., 1966, *J. Geophys. Res.* **71**, 4495.

Dessler, A. J., 1958, *J. Geophys. Res.* **63**, 507.

Dessler, A. J., and Parker, E. N., 1959, *J. Geophys. Res.* **64**, 2239.

Dungey, J. W., 1955, *Physics of the Ionosphere,* Physical Society, London.

Eather, R. H., 1967, *Rev. Geophys.* **5**, 207.

Fejer, J. A., 1961, *Canadian J. Phys.* **39**, 1409.

Feldstein, Y. I., 1963, *Geomagnetism and Aeronomy* **3**, 183.

Fukushima, N., 1953, *J. Fac. Sci. Univ. Tokyo,* Sect. II, Part V, **8**, 291.

Gurnett, D. A. and O'Brien, B. J., 1964, *J. Geophys. Res.* **69**, 65.

Hartz, T. R. and Brice, N. M., 1967, *Planet. Space Sci.* **15**, 301,

Hines, C. O., 1964, *Space Sci. Rev.* **3**, 342.

Kato, Y. and Watanabe, Y., 1955, *Sci. Rep. Tohoku Univ. Ser.* 5, **6**, 95.

Kennel, C. F. and Petschek, H. E., 1966, *J. Geophys. Res.* **71**, 1.

Lehnert, B., 1956, *Tellus* **8**, 241.

Matsushita, S. and Campbell, W. H., eds., 1967, *Physics of Geomagnetic Phenomena,* Academic Press, New York.

McCormac, B. M., 1968, *Earth's Particles and Fields,* Reinhold, New York.

McIlwain, C. E., 1961, *J. Geophys. Res.* **66**, 3681.

Mead, G. C., 1964, *J. Geophys. Res.* **69**, 1181.

Milton, D. W., McPherron, R. L., Anderson, K. A. and Ward, S. H., 1967, *J. Geophys. Res.* **72**, 414.

Nagata, T., 1963, *Planet. Space Sci.* **11**, 1395.

Nagata, T. and Kokubun, S., 1962, *Rep. Ionosphere Space Res. Japan* **16**, 256.

Nishida, A., 1968, *J. Geophys. Res.* **73**, 1795.

Northrop, T. G., 1963, *The Adiabatic Motion of Charged Particles,* Interscience Publishers, New York.

Obayashi, T., 1964, *Space Sci. Rev.* **3**, 79.

Obayashi, T., 1967, *Solar Terrestrial Physics,* J. W. King and W. S. Newman, eds., Academic Press, London.

O'Brien, B. J., 1964, *J. Geophys. Res.* **69**, 13.

O'Brien, B. J. and Taylor, H., 1964, *J. Geophys. Res.* **69**, 45.

Parker, E. N., 1962, *Space Sci. Rev.* **1**, 62.

Parks, G. K., Coroniti, F. V., McPherron, R. L. and Anderson, K. A., 1968, *J. Geophys. Res.* **73**, 1685, 1697 and 1715.

Piddington, J. H., 1959, *Geophys. J. Roy. Astron. Soc.* **2**, 173.

Piddington, J. H., 1960, *Geophys. J. Roy. Astron. Soc.* **3**, 314.

Piddington, J. H., 1964, *Space Sci. Rev.* **3**, 724.

Piddington, J. H., 1965, *Planet. Space Sci.* **13**, 565.

Roach, F. E. and Roach, J. R., 1963, *Planet. Space Sci.* **11**, 523.

Sandford, B. P., 1964, *J. Atmos. Terr. Phys.* **26**, 749.

Sandford, B. P., 1966, *New Zealand Dept. Sci. Indust. Res. No. R329.*

Scarf, F. L., Fredericks, R. W. and Crook, G. M., 1968, *J. Geophys. Res.* **73**, 1723.

Shabanski, V. P., 1968, *Space Sci. Rev.* **8**, 366.

Singer, S. F., 1957, *Trans. American Geophys. Union* **38**, 175.

Störmer, C., 1955, *The Polar Aurora,* Oxford University Press, London.

Sugiura, M. and Chapman, S., 1960, *Abh. Akad. Wiss. Göttingen, Math. Phys. Kl.* Sonderheft 4.

Winckler, J. R., Bhavsar, P. D. and Anderson, K. A., 1962, *J. Geophys. Res.* **67**, 3717.

CHAPTER 9

Theory of the Earth's Radiation Belt, Aurora and Ionospheric Currents

In Chapter 6 we saw how the solar wind compressed the geomagnetic field to form a cavity with a boundary some 10 R_E along the earth-sun line. The region between the ionosphere and this boundary is called the magnetosphere. In Chapter 7 the magnetotail was described and shown to be the storehouse of a large and variable amount of magnetic energy. This could be transformed to particle energy and fed back into the corotating magnetosphere to power the Van Allen belt, auroras and ionospheric current systems. In Chapter 8 the mechanism of these phenomena was discussed. In its simplest form it involved rapid magnetic field reconnection across the tail neutral sheet (Figure 7.1) and a flow of fast particles back into the corotating magnetosphere, where some were precipitated (auroras) and some inflated the magneto-sphere (radiation belt). A pattern of convective flow of plasma and frozen-in field lines was described (Figure 8.9) involving motions to and from the tail and through the body of the corotating magneto-sphere. The corresponding motions of the feet of the field lines (in the ionosphere) and their attached electrons might give rise to the DP2 current system (Figure 8.4).

In this chapter we consider some more recent, and more speculative, electrodynamic theory of the earth's magnetosphere. The various phenomena are extremely complex and are far from being understood. However, one approach which is relatively simple, yet revealing, is to determine energy requirements and energy sources.

The total energy requirement for a great storm is shown in Section 9.1 to exceed 10^{23} erg and possibly to exceed 10^{24} erg. The original source of the energy must be the solar wind and the energy transfer must occur, not on the solar side of the magnetosphere, but rather on the sides and along the tail surface. It was seen that the magnetic energy stored in the tail out to 100 R_E must exceed 10^{23} erg, and will be even more if we take into account more distant parts of the tail.

176

However, even this estimate may be much lower than the total available storm energy because energy flows into the tail and out of the tail continuously, and the amount stored at a given instant may be much less than the total input.

To determine the total input we must set up tail models based on Figure 7.7 or Figure 7.8; it does not matter which but the latter is easier to visualize. Suppose that at a distance x along the tail the cross section is a circle of radius $R(x)$, and the field strengths are $B(x)$, $B_n(x)$ along the tail, across the neutral sheet and $B_s(x)$ at the outer surface. The magnetic flux into each section is $(\pi R^2/2)B$ and this decreases with x since flux $2RB_n$ per unit length connects across the neutral sheet, and flux πRB_s per unit length connects with the interplanetary field. Hence we have

$$\frac{\partial B}{\partial x} = -\frac{2}{\pi R}(2B_n + \pi B_s) \tag{9.1}$$

Assuming R constant, this equation could be used to find B_n or B_s if either one of these small quantities could be measured.

Here it may be used to show how large and variable may be the power input to the tail. Suppose, for example, that reconnection through the neutral sheet is stopped for a period sufficient for all lines already reconnected to return to the corotating magnetosphere. This period is an hour, or a few hours, and subsequently we have $B_n = 0$. At the same time lines continue to move into the tail and so we have a growing tail. Assume, for simplicity, that B_s is a constant over the full length X of the tail so that equation 9.1 may be integrated to give

$$B = B_o - \frac{2}{R}B_s x \qquad\qquad B_o = \frac{2}{R}B_s X \tag{9.2}$$

where $B_o = B(0)$. A tail of length 20 R (about 400 R_E) has $B_s = B_o/40$ (about 0.5 γ). The tangential (frictional) force exerted by the solar wind is $f = BB_s/4\pi$, and if the wind velocity is V, then the power input is

$$W = \int 2\pi R f V dx \tag{9.3}$$

Substituting from equation 9.2 and integrating, we have

$$W = \frac{1}{8}B_o{}^2 R^2 V \tag{9.4}$$

an equation which is obtainable more easily by multiplying the tail tension per unit area $(B_o{}^2/8\pi)$ by the area (πR^2) and the rate of extension (V). As we have seen, the storm-time values of the parameters are

$B_o \sim 40\ \gamma$, $V \sim 700$ km/sec, $R \sim 20\ R_E$ so that $W \sim 2 \times 10^{19}$ erg/sec. This power may be compared with the total power of the solar wind ($N = 10/\text{cm}^3$, $V = 700$ km/sec) incident on the front of the corotating magnetosphere (radius $10\ R_L$); this is about 4×10^{20} erg/sec. Thus the power fed to the tail may be much more than the usually assumed "tiny fraction" of the frontally intercepted power.

On the other hand if all, or nearly all, of the tail flux reconnects across the neutral sheet so that $B_s \sim 0$, then little or no power is being fed to the tail. We then have a *decaying tail*. The growth or decay of the magnetotail will depend more or less equally on two factors: the rate of flow of field lines into the tail and the rate of flow back from the tail. At first sight it would seem that increased flow to the tail is the effect most likely to lead to a storm, but this is not necessarily so. Inhibition of flow from the tail has two effects: it causes tail growth and it also increases the distance to the region of reconnection. In Figure 7.7 field line 4 has reconnected at the dot and only the magnetic energy between the dot and the corotating magnetosphere is available as storm energy. The distance to the reconnection point is also all-important in determining the amount of plasma drawn back into the corotating magnetosphere to provide the storm-time radiation belt. Thus it may be temporarily decreased flow from the tail that is the principal cause of some storms or substorms.

9.1   The Storm-Time Radiation Belt

The early theory of the geomagnetic tail (Chapter 7) included the concept of field lines reconnecting across the neutral sheet and contracting elastically back to the corotating magnetosphere as shown in Figure 7.1. One may explain this plasma acceleration in terms of a Lorentz force $\mathbf{j} \times \mathbf{B}$, where $\mathbf{j}$ is the density of the dawn-to-dusk tail current sheet and $\mathbf{B}$ is the northward field across the sheet. Acceleration continues until the field is destroyed and if the field strength and plasma particle density in a part of the tail are $20\ \gamma$ and $1/\text{cm}^3$, respectively, then when magnetic energy is converted to particle kinetic energy the latter will average about 10^3 eV.

It is convenient to divide the incoming plasma particles into two groups: those with generally small pitch angles and those with large pitch angles. The former move along the field lines and bounce from hemisphere to hemisphere in the manner discussed in Section 8.4 (although their journey, which takes them down the tail and across the neutral sheet, is much longer than that of belt particles). These particles experience first-order Fermi acceleration (Sakurai, 1966, and Section 4.4) and are potential auroral particles; their fate is discussed in Section 9.4. The second group of particles, with large pitch angles, remains in or near the neutral sheet. They are drawn towards the earth, and as the field strength B increases they experience betatron acceleration perpendicular to the field direction. The two particle groups remain largely separate because of the very low collision frequency.

The second group remains firmly trapped and shares the general convective motion V_q which was illustrated in Figure 8.9. This motion carries the hot plasma into the rear of the corotating magnetosphere and if this were the only motion, it would carry it out through the front and back to the tail as shown. However, it is not the only motion and much of the fresh, hot plasma remains in the corotating magnetosphere to form the storm-time radiation belt and ring current. The situation is shown schematically in Figure 9.1, which is an equatorial section of the magnetosphere and environs according to the particle distribution. The bow wave and magnetosphere boundary enclose a hatched region called the "skirt" in which particle and field parameters vary rapidly, and through which there may be hydromagnetic flow of plasma and field lines to the tail. The central region is the zone of stable trapping and in its rear is shown a storm-time radiation belt beginning to form. The "cusp" corresponds to the region between the tail and the trapping region and it is through the cusp that fresh hot plasma is introduced by hydromagnetic flow as shown by the arrows.

In moving from the tail neutral sheet to the trapping region a particle may experience an increase in magnetic field strength from 1 γ to 100 γ, and a betatron acceleration from 1 keV to 100 keV. Inside the corotating magnetosphere the particle experiences magnetic field gradients, not only of the dipole field, but of the disturbance caused by the presence of the inflating plasma itself. The principal effect is a westward drift of protons and an eastward drift of electrons, and this results in the development of electric space charge as shown in Figure 9.1. The effects of this space charge are discussed below, but meanwhile it is evident that the combination of the various drifts, together with the earth's rotation which presents a changing aspect of the corotating magnetosphere to the tail, must spread some of the fresh plasma into a complete radiation belt.

The above mechanism explains qualitatively the main observed features of the storm-time radiation belt, which seems to develop first in the dark hemisphere and spread around the earth in a few hours. One important effect of this radiation belt is that it constitutes a ring current, and causes the main-phase (D_{st}) decrease illustrated in Figure

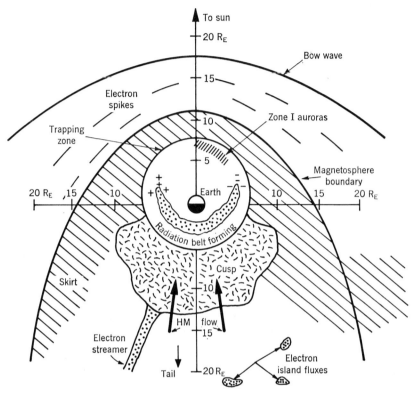

Fig. 9.1. An equatorial cross section of the magnetosphere inferred from magnetic field measurements and spatial distributions of energetic particles (mainly electrons of energy > 40 keV). An inner zone of stable trapping extends during quiet periods to about 8 R_E towards the sun and 6 R_E away from the sun. Within this region are shown the radiation belt forming in the dark hemisphere and the Zone I auroral region on the day side. At the front and sides of the inner zone is the "skirt," of unstable trapping and rapid particle and field fluctuations. At the rear is the "cusp" and then the tail. Electron "island fluxes" and "spikes" are found in the tail plasma sheet and magnetosheath, respectively.

8.1. Theory shows that the amplitude of this decrease is proportional to the total particle energy in the belt and that a decrease of 100 γ requires a belt of energy a few times 10^{22} erg. Observed decreases for great storms range up to about 500 γ and, as Akasofu and Yoshida (1966) have pointed out, losses from the belt may be heavy during its period of development. When allowances are made for such storms and such losses the total energy requirement may exceed 10^{24} erg and the total power requirement may exceed 10^{19} erg/sec. As we have seen, this energy requirement may be met from the tail, provided reconnection takes place at a distance of a few hundred R_E down the tail. There does not seem to be any alternative source.

Evidence in favor of the above belt model is provided by the observation by Konradi (1967) of the sudden appearance of protons of energy a few times 10^5 eV in the front of the magnetosphere. These appeared as a storm developed, and more energetic protons appeared first. It was possible to show that all time differences were consistent with the introduction of the protons in the rear of the magnetosphere, and their drift under the influence of the gradient of the geomagnetic field.

9.2 The Magnetotail Plasma Sheet

The convective motions V_q of magnetic field lines and plasma to and from the magnetotail seem to account for the development of the radiation belt, as described above. The radiation belt accounts for the main-phase decrease D_{st} of the geomagnetic storm. As seen in Section 8.5, the convective motions, transferred along the magnetic field lines to the ionosphere, account for the DP2 current system. They also appear to account for some auroral precipitation, and until 1967 it was widely believed that this convective pattern was basic to all storm phenomena, which waxed and waned with the rate of flow.

It is now believed that this model accounts for only some of the various storm-time effects, and that during a substorm the whole pattern of convection changes. In Figure 8.1 the story of a geomagnetic storm is shown schematically and it is seen that during the first day a series of magnetic substorms occurs, lasting an hour or so. These are also called negative bays, and their ionospheric current system is that described in Section 8.1 as the DP1 system. The magnetic substorm is part of a magnetospheric substorm and is accompanied by an auroral substorm and other effects. It may occur as one of a series of substorms as in Figure 8.1, or as an isolated event. According to the concept of hydromagnetic flow from the magnetotail, one would expect a substorm, whatever its precise form, to follow after a disturbance in the tail. However, when spacecraft began to make regular measurements in the tail it was soon found that some particle and field disturbances occurred in the tail after the observation of substorms at sea level.

One such effect is a change in magnetic field strength in the tail (Heppner et al., 1967). Corresponding particle events were observed by Hones et al. (1967) and one type of event is shown in Figure 9.2. At the commencement of a substorm or negative bay, shown as a plot of the X component of the geomagnetic field at Kiruna Observatory, the flux P of energetic electrons (> 45 keV) at a distance of about 17 R_E

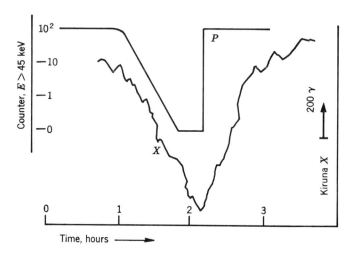

Fig. 9.2. A plot of a geomagnetic "bay" (the X component of the geomagnetic field at Kiruna) and the corresponding slow disappearance and abrupt appearance of the tail plasma sheet (after Hones *et al.*, 1967).

down the tail gradually decreased to zero. As the bay attained its maximum intensity the flux abruptly increased to its earlier value and then remained steady. Hones *et al.* interpreted their results as a squeezing out of the plasma from the magnetic flux tubes, caused by increased magnetic pressure.

In addition to these temporary disappearances of the tail plasma sheet, they also observed many abrupt increases in particle flux from zero to a level comparable with that in the plasma sheet itself. The increases would be followed usually by a slow decay, and the event associated with a negative bay. It would seem that these events must be identified with the "islands" of electrons and subnormal magnetic field strength observed and illustrated in Figure 9.1. If this is the case, then the "islands" are misnamed, and are really "streamers" of fast electrons stretching out from the cusp as shown on the left of Figure 9.1. It is likely that the same phenomenon accounts for the "spikes" of electrons observed just above the ionosphere by Fritz and Gurnett (1965) and others and illustrated in Figure 7.4. These also occur on magnetic force tubes of limited cross section, which project into the tail. A most significant feature of the tail "streamers" is that their electrons are moving away from the earth (Rothwell, 1967). A similar conclusion was reached by Armstrong and Krimigis (1968), who observed energetic protons (> 0.3 MeV) in the tail at distances of 80 R_E.

These also streamed away from the earth and were well correlated with negative bays.

All of these observations show that fast particles are ejected from the corotating magnetosphere or cusp into the tail. They also suggest that the tail plasma sheet may, at least on some occasions, develop as a result of outward movement of hot plasma. The situation is shown schematically in Figure 9.3, with the plasma sheet thickness some

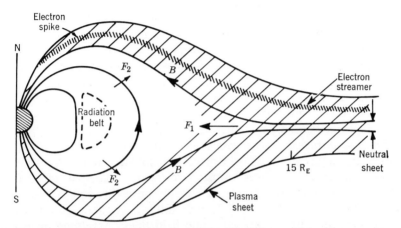

Fig. 9.3. A section of the tail plasma sheet in the plane of the midnight meridian, and extending out to about 20 R_E. The magnetic neutral sheet (0.1–1 R_E thick), the hatched plasma sheet (2–6 R_E thick) and the radiation belt are shown, but the magnetic tail outside this region is not shown.

twenty-five times that of the magnetic neutral sheet. It is seen that a tail "streamer" and electron "spike" may occur far from the magnetic neutral sheet, and so could not be energized from that sheet.

It would seem that there are two conflicting pictures of plasma flow to and from the tail. The first is hydromagnetic flow of plasma and field lines back from the tail, following reconnection across the neutral sheet. This is shown by the arrow marked F_1, and as we have seen is strongly supported by energy considerations because there is no alternative to tail magnetic energy. The second flow pattern F_2 is supported by the above observational data and by theoretical arguments given briefly below. The two flow patterns are found to be consistent and the change from F_1 to F_2 results in a change in the ionospheric current system from DP2 to DP1 (Piddington, 1968).

The outward flow F_2 is thought to be one manifestation of the magnetospheric substorm. An increase in the inward flow F_1 during the first day of a storm develops the radiation belt, strongest in the night sector. This asymmetric belt and its interactions with the magnetosphere and ionosphere provide an incredibly complex series of effects, one of which is the ejection of plasma into the tail. The drift motion of a particle in the equatorial plane is given by equation (see equation 8.8),

$$\mathbf{u}_\perp = -\frac{\mathbf{B}}{B^2} \times (\mathbf{E} - \frac{\mu}{e}\nabla B) \qquad (9.5)$$

The drift due to the centrifugal force on the particle is ignored, and for the time being the magnetic field gradient is taken as that of a dipole geomagnetic field. This gradient causes a westward drift of 40 keV protons of a few radians per hour, which is much faster than the convection (V_q of Figure 8.9) of the ambient plasma of about 0.5 radian/hour. Electron drift is eastward and so a symmetrical radiation belt provides a ring current, but an asymmetrical belt becomes electrically polarized as shown in Figure 9.1.

This asymmetric ring current is connected to the ionosphere as shown in Figure 8.8, and will drive ionospheric Hall current cells as shown there. It also drives ionospheric Pedersen current and it is this effect in which we are interested here because this causes belt losses. We assume that the radiation belt and ionosphere are in perfect electrical connection along magnetic field lines. Assuming an average night-time value of ionospheric conductivity, it is found that a radiation belt of energy density 10^{-8} erg/cm^3 is able to supply the ionospheric losses corresponding to a magnetospheric electric field (across the field lines) of about 3×10^3 emu. The energy density of protons in the storm-time radiation belt has been measured by Frank (1967) and it does attain the above value.

The resulting electric field $\mathbf{E}$ gives rise to Hall drift also according to equation 9.5; this drift is away from the earth and into the tail. For a magnetic field strength of 200 γ and the above electric field, the drift rate is about 10 R_E/hour, so that within the time scale of the substorm the plasma is ejected back into the tail. The drift is, in effect, an un-freezing of the plasma and field lines. The plasma moves so as to invade tubes of force at higher latitudes and so as to recreate the tail plasma sheet which had been swept inwards by the earlier accelerated flow F_1 of Figure 9.3. The new flow pattern is shown as F_2 in Figure 9.3 and as V_s in Figure 8.9. It may account for the sudden reappearance of the tail plasma sheet as in Figure 9.2. Concentrations of hot plasma of smaller dimensions than the half radiation belt will also become polarized and these may be responsible for the "islands" or "streamers" of electrons shown in Figures 9.1 and 9.3.

The above discussion may provide a qualitative explanation of the tail plasma sheet, but it is greatly simplified. A satisfactory quantitative discussion must also include the following:

(1) The effects of the ∇B and **E** drifts (equation 9.5) operating simultaneously.

(2) The interaction between the belt polarization **E** and ionospheric Pedersen and Hall currents, including the effects of ionosphere polarization fields.

(3) The influence on ∇B of the ring-current plasma itself. Replacement of magnetic pressure by plasma pressure will change the radial gradient and introduce azimuthal gradients of field strength.

(4) The likelihood of development of instabilities, notably the interchange instability and the Rayleigh-Taylor instability in the rear of the corotating magnetosphere. This possibility is not separable from those listed as (1)–(3), but rather includes them. The interchange instability must include several different factors and these have been discussed by Swift (1964, 1967) and by Chang *et al.* (1965). The Rayleigh-Taylor instability does not seem to have been considered, although it may be required to explain the limited cross section (in both latitude and longitude) of the tail "islands" or "streamers."

9.3 Ionospheric Current Systems

Apart from the tidal dynamo or Sq current systems (Section 3.5), ionospheric current systems are driven by electric fields originating in the magnetosphere. A simple way of regarding these fields is that they result from the convective motions of magnetospheric plasma and field lines; the feet of the latter move through the stationary ionospheric gas to provide an electric field $E = -V \times B$. It seems that the drift **V** and field **E** have two quite different forms: DP2 for average conditions and DP1 during substorms. A review and discussion of these drifts and their electric fields has been given by Obayashi and Nishida (1968).

Again we must stress the extreme difficulty in understanding ionospheric current systems. In the first place they are inferred mainly from a limited number of sea-level magnetic measurements. There is not even complete agreement on the forms of the equivalent current systems, let alone the real systems. The problem is then compounded by our lack of knowledge of ionospheric conductivity, particularly at high latitudes and during the night. The above explanation of the DP2 current system in terms of the drift V_q may eventually require review.

However, it seems to make the best use of available data and theory and it probably includes the principal effects.

The DP1 current system may have one or more of the forms shown in Figures 8.2 and 8.3. It is responsible for the magnetic substorms shown schematically in Figure 8.1, which are part of the DS system. They are shown on a much reduced scale, and at auroral latitudes the sea-level perturbations may exceed 1000 γ. Early theories of this current system have been reviewed by Akasofu (1966), but none may be developed with any confidence when the form of the current system itself is in doubt. However, it is fairly certain that two principal factors are involved in the change from DP2 to DP1 as a storm develops. The first is a change in the large-scale magnetospheric electric field and the second is an increase by a factor of 10 to 1000 in the ionospheric conductivity along an auroral arc or band.

A comparison of Figures 8.2 and 8.4 shows that each has a pair of current cells centered at high latitudes ($\gtrsim 70°$), which combine to send current over the polar cap from the night side to the day side. In the DP1 system there are two additional current cells centered near latitude 53°. These squeeze the polar cells into latitudes above the auroral oval, and between each of the adjoining cell pairs appears a region of high current density called an electrojet. These current systems may be fairly well accounted for in terms of an asymmetric ring current on the night side as shown in Section 8.5. Some improvement seems possible by taking into account variations in ionospheric conductivity measured horizontally (Gottlieb and Fejer, 1967). A further important advance is possible by taking account of the feedback between the ionosphere and the magnetosphere (Swift, 1968). The electric field in the latter is transferred along the field lines to the ionosphere where a complicated current system is set up. The requirement that the whole, three-dimensional current system is divergenceless requires that electric space-charge fields are set up in the ionosphere. These are transferred back to the magnetosphere where they affect the motions of the radiation belt particles and modify the original space-charge fields. The whole process must be examined in a self-consistent way.

At this stage the theory of the DP1 current system has become an order of magnitude more difficult than that of the tidal dynamo (Sq) current system, which itself is still incomplete because of the complexity of the ionospheric conductivity distribution. However, Swift's theory still involves several approximations which (as he points out) leave it far from complete. In the first place his solution is for a steady state which might take as long as a day to attain, whereas notable changes in some parameters occur within an hour or less (Figure 9.2). Second, his theory is linear, applying only to small perturbations and so not taking account of the plasma stresses on the magnetospheric field. Third, his assumed ionospheric conductivity in the auroral zone exceeds the average by a factor of only ten, whereas observational data include much larger and highly variable ratios. Clearly the inclusion of these additional factors would make the theory excessively complex, and although it has been most illuminating, further development along these lines may not be desirable, particularly as the results are quite sensitive to changes in ionospheric conductivity and to plasma density gradients.

Other forms of the DP1 current system which have been suggested are shown in Figure 8.3. In view of the above discussion it would seem that a wide variation of these systems may indeed occur, depending on the magnetospheric plasma distribution and pattern of ionospheric conductivity. The notable feature of these two systems is the westward electrojet, which gives rise to the negative bay at nearby sea-level observatories (Figure 9.2). Another approach to the DP1 theory is to concentrate on this electrojet; if it can be explained, then the remainder is simply "return currents." One approach is to identify the electrojet with the auroral arc in which the vertically integrated Hall conductivity ($\Sigma_2 \sim 6 \times 10^{-8}$ emu) is about 300 times that of the surrounding ionosphere and the Pederson conductivity ($\Sigma_1 \sim 4 \times 10^{-8}$ emu) about 60 times that of its surroundings (Boström, 1964). The second requirement is an electric field along the auroral arc; the electrojet is then provided by the Cowling conductivity (Section 3.3) which is larger again than Σ_2.

There seem to be at least two ways in which a westward electric field might be introduced along the night side of the auroral oval.

(1) The convection pattern V_q (Figures 8.4 and 8.9) provides an electric field $E_q = -V_q \times B$ which, when imposed on a normal ionosphere, may provide the DP2 current system. When imposed on the highly conducting auroral arc it may provide the DP1 current system in the following manner. The motion V_q in the ionosphere is a southward drift (in the northern hemisphere) across the auroral arc. The moving field lines carry the electrons southward, but not the ions which are held back by collisions. The result is a southward electric field and a very intense westward Hall current (which is the Cowling current corresponding to E_q). This electrojet mechanism has been discussed quantitatively by Atkinson (1967).

(2) An alternative, or additional, cause of a westward electric field is the interplanetary electric field (Section 7.3). If this is able to penetrate

between the tail sections and down to the ionosphere it would provide the negative bay and explain the close correlation between this phenomenon and a southward directed interplanetary magnetic field.

Summing up, it seems that the ionospheric *DS* current system is highly variable in form as well as intensity. It depends very much on changing distributions of ionospheric conductivity and is driven by combinations of at least two and perhaps more electric field systems.

9.4   Auroras

Most auroral emission results from the bombardment of the upper atmosphere by energetic particles moving down from the magnetosphere. Such precipitation depends on the plasma distribution and on the electromagnetic field in the magnetosphere, and as seen above these are complex, highly variable and not well understood. Naturally many attempts have been made to explain auroral precipitation and these have been reviewed by Akasofu (1966). Since the electric field system in the magnetosphere is not understood, it follows that no complete precipitation theory has been developed, although some of the effects discussed are undoubtedly important. For example, the electric field caused by the asymmetrical ring current must have notable effects, as must hydromagnetic flow from the tail.

In discussing more recent auroral theory it is useful to divide auroras into the various types described in Section 8.2, notably Zone II (discrete, visual auroras) and Zone I (diffuse, mantle auroras).

Discrete, Visual Auroras

Discrete, visual auroras have received by far the most attention of all auroral types, and until relatively recently these were the only recognized "auroras." The energy required to maintain these auroras during storm times probably exceeds 10^{18} erg/sec, and the only apparent source seems to be the tail. As described above, field lines which have reconnected in the distant tail contract towards the earth and provide an adequate energy source. Particles with rather small pitch angles experience first-order Fermi acceleration along the field lines and some are "dumped" into the atmosphere along the night sector of the auroral oval. Particles which remain trapped enter the corotating magnetosphere to form the radiation belt, but some of these may later be expelled and in passing outwards through the cusp region (Figure 9.1) may be precipitated or may be ejected into the tail.

This model may indicate the general source of discrete auroral particles, but it is clearly far from providing a full theory. In the first place auroral arcs and bands often occur in multiple form as shown in Figure 8.5. Arc or band thickness and separation may be a few hundred meters and a few tens of kilometers, respectively, a ratio of about 100. The model would provide a single, broad arc or band in the dark hemisphere whereas complex forms are observed in both hemispheres (Figure 8.6).

One effect which is almost certainly of great importance is the generation of electric space-charge fields as a result of the interaction of magnetospheric convection V_s with ionospheric irregularities. As shown above a southward convection across an east-west auroral arc (of high ionization density) will give rise to a southward electric field. The space charge also projects electric fields into the magnetosphere, upwards on the northern side of the arc and downwards on the southern side, and Atkinson (1968) has suggested that the upward field increases the rate of auroral electron precipitation and the ionization density, and so provides a regenerative process. An alternative, and I believe more likely, effect is that the weak vertical fields cause motion of the low-energy ambient electrons. This results in an upward electric current on the north side and a downward current on the south. The magnetic field lines through the arc gain an eastward component and become tilted, and the magnetic shell projecting from the arc becomes a standing *hydromagnetic shear wave*. It is this wave which may enhance precipitation of high-energy electrons.

Another effect which may be important in connection with auroras, but which has received little attention, is convergence of existing ionospheric neutral ionization (equal numbers of electrons and positive ions). A simple model illustrating this effect is shown in Figure 3.2 and others have been discussed (Piddington, 1963) in connection with auroras. Such convergence increases ionization density and so increases auroral emission as well as electrical conductivity.

Finally, we refer briefly to daytime Zone II auroras (Figure 8.6), which may be a result of the formation of magnetic neutral sheets on the solar side of the magnetosphere, or on the outer surface of the tail (Figure 7.1).

Diffuse, Mantle Auroras

Zone I, or diffuse, auroras have many different characteristics to those discussed above, and must have a different origin. Some theories which have been developed to explain auroras in general, without specifying the particular type, seem to fail when applied to Zone II (discrete) auroras but may be valuable in connection with the diffuse type. We refer to the various attempts to determine a magnetospheric electric field system, taking account of an asymmetric radiation belt and one of a variety of possible ionospheric conductivity distributions (Swift, 1968).

Another approach to this whole problem, which is rather less ambitious, is to accept a particular ionospheric current system as a starting point. Taylor and Hones (1965) assumed that the statistical, equivalent *DS* current system (rather similar to that of Figure 8.2) was a real current system. Assuming values of Pedersen and Hall conductivity they then determined the electric field required to drive this current, and projected this field into the magnetosphere. The motions of energetic electrons and protons were determined by assuming the first two adiabatic invariants conserved and the total particle energy (kinetic plus electric potential) constant. Regions where particle motions took them into the atmosphere were then plotted as auroral zones. This approach seems to offer real possibilities of explaining Zone I auroras, but it meets two major difficulties. The real ionospheric current system may differ greatly from the equivalent current system used; indeed, as we have seen, the latter may be mainly due to Hall current with the sea-level effects of Pedersen currents cancelling out. Second, the ionospheric conductivity, particularly at night, is not known, and at high latitudes it is highly variable. Some variation is caused by precipitation and here again we have the probability of two-way interaction between the trapped particles and the ionosphere which sometimes absorbs them.

The physics of auroras is indeed complex, and its understanding must develop in parallel with that of ionospheric current systems, Van Allen radiation and the electromagnetic field in the magnetosphere. Together these provide a rich field for future study.

References

Akasofu, S.-I., 1966, *Space Sci. Rev.* **6**, 21.
Akasofu, S.-I. and Yoshida, S., 1966, *J. Geophys Res.* **71**, 231.
Armstrong, T. P. and Krimigis, S. M., 1968, *J. Geophys. Res.* **73**, 143.
Atkinson, G., 1967, *J. Geophys. Res.* **72**, 5373 and 6063.
Atkinson, G., 1968, "Auroral Arcs, Result of, etc.," preprint, Jet Propulsion Laboratory.
Boström, R., 1964, *J. Geophys. Res.* **69**, 4983.
Chang, D. B., Pearlstein, L. D. and Rosenbluth, M. N., 1965, *J. Geophys. Res.* **70**, 3085.

Frank, L. A., 1967, *J. Geophys. Res.* **72**, 1905 and 3753.

Fritz, T. A. and Gurnett, D. A., 1965, *J. Geophys. Res.* **70**, 2485.

Gottlieb, B. and Fejer, J. A., 1967, *J. Geophys. Res.* **72**, 239.

Heppner, S. P., Suguira, M., Skillman, T. L. and Ledley, B. G., 1967, *J. Geophys. Res.* **72**, 5417.

Hones, E. W., Asbridge, J. R., Bame, S. J. and Strong, I. B., 1967, *J. Geophys. Res.* **72**, 5879.

Konradi, A., 1967, *J. Geophys. Res.* **72**, 3829.

Obayashi, T. and Nishida, A., 1968, *Space Sci. Rev.* **8**, 3.

Piddington, J. H., 1963, *Geophys. J. Roy. Astron. Soc.* **7**, 415.

Piddington, J. H., 1968, *Planet. Space Sci.* **16**, 703.

Rothwell, P., 1967, *Space Sci. Rev.* **7**, 278.

Sakurai, K., 1966, *Rep. Ionospheric Space Res. Japan* **20**, 49.

Swift, D. W., 1964, *Planet. Space Sci.* **12**, 945.

Swift, D. W., 1967, *Planet. Space Sci.* **15**, 835 and 1225.

Swift, D. W., 1968, *Planet. Space Sci.* **16**, 329.

Taylor, H. E. and Hones, E. W., 1965, *J. Geophys. Res.* **70**, 3605.

CHAPTER 10

Planets, Satellites and Comets

Our seven major planetary neighbors and some of their satellites, as well as the cometary interlopers, have been studied intensively by their optical and radio emissions, by radar techniques and a few by direct space probes. The radio emissions, caused by the acceleration of electrons in a variety of electric and magnetic fields, have been particularly useful in revealing interesting electrodynamic effects. These effects are proving important in a wide variety of studies ranging from the origin of the solar system itself to the orientation of comet tails and to the strange control of one of Jupiter's satellites over the radio emission from the planet.

Radio studies of the moon and planets have long provided measurements of the thermal emissions from the surface layer and atmosphere of these objects and so allowed estimates to be made of temperatures. Radar studies allow estimates of surface roughness and rates of rotation, and will be used as a test of the theory of general relativity by the effect of solar gravity on the delay times of radar echoes from Mercury and Venus. Earlier studies are discussed in various books on radio astronomy and the more recent by Mayer and Spangler (1967).

Four more recent discoveries illustrate the amazing diversity of information revealed by radio observations. First, the inference from radio data that Mercury and Venus have peculiar rotation patterns, the first non-synchronous with a 59-day period and the second retrograde (243 days). Second, the observational identification of the 10 cm radio emission from Venus with the solid surface. Third, the very successful exploitation of a new method of investigating planets and their atmospheres by observing the refraction and occultation of radio signals from spacecraft which move behind the planet. Fourth, the observation for the first time of radio emission from the planets Uranus and Neptune.

In this chapter we consider some of the more interesting electrodynamical problems of the solar system, the first being its origin.

192

Attempts to explain its origin date back more than three hundred years, but were generally unsuccessful until account was taken of magnetic fields and their effects, which are described in Section 10.1.

Perhaps the most interesting planet from the point of view of observed electrodynamic effects is Jupiter. It has a powerful and presumably very extensive magnetic field which, taken in conjunction with a rapid rate of rotation, raises some interesting theoretical problems involving magnetic interchange motions and frictional interaction with the solar wind. It also possesses a Van Allen type of radiation belt which presumably requires a different explanation to that of the earth's belt. Most remarkable of all is the control of its radio emissions by one of its twelve satellites (Io). These various phenomena are discussed in Sections 10.2 and 10.3.

Other planets which provide interesting electrodynamic problems are Venus, Mars and Saturn. Our two nearest neighbors are of interest mainly because space probes have made measurements in their vicinities and measured particle fluxes and magnetic fields. Saturn is potentially interesting because of some hydromagnetic interactions between the ring and the atmosphere which must occur if the planet has a substantial magnetic field. These various effects are discussed in Section 10.4.

The moon has proven a rather disappointing object in a number of ways. Not only does it lack a magnetic field of internal origin, but it has such low electrical conductivity (at least in the surface layer) that it interferes only slightly with the interplanetary field. Nevertheless an extensive series of papers on lunar electrodynamics has appeared, one possible incentive for which is the likelihood that man-made measurements of lunar fields and particles will be possible. These papers are reviewed in Section 10.5.

Finally, in Section 10.6 comets are discussed; they are of interest here because of their hydromagnetic interaction with the solar wind, which provided some of the earliest evidence of the existence of the wind.

10.1 The Origin of the Solar System

The origin of the solar system is one of the oldest unsolved questions of natural philosophy, and even relatively modern discussions date back to Descartes, Kant and Laplace in the seventeenth and eighteenth centuries (for references see reviews by Williams and Cremin, 1968, ter Haar, 1967 or Jastrow and Cameron, 1963). However, until relatively recently very little progress was made, and Jeffreys concluded the Bakerian lecture of 1952 with the remark that "there is hardly a feature of our system that I would regard as satisfactorily explained."

Since then considerable progress has been made as a result of developments in three different fields. First, the subject of cosmic electrodynamics has made great advances and promises to play a major part in our understanding of the solar system. Second, we have the advances in space exploration which have extended to the moon and the two nearest planets and promise to reveal past history of these and other members of the solar system. Third, a great deal of important information has been secured from the investigation of meteorites.

Before considering theories of the origin of the solar system, it is useful to summarize its major properties.

(1) The central condensation, the sun, is more massive by a factor of about 10^3 than all of the remaining parts of the system.

(2) The angular momentum of the sun is about 1% that of the planets.

(3) The vectors associated with the angular momenta of nearly all of the parts of the system are approximately parallel.

(4) Of the eight planets (regarding Pluto as an escaped satellite), the four nearest the sun (the terrestrial planets) are less massive than the major planets by a factor of about 100.

(5) The terrestrial planets are composed predominantly of heavy non-volatile material; the major planets are composed of hydrogen and helium, the raw material from which the sun was formed.

(6) An empirical law giving the planetary distances in astronomical units may be written

$$r = 0.4 + 0.3 \times 2^n \qquad (10.1)$$

where $n = -\infty$ for Mercury, 0 for Venus, 1 for the earth and so on. This is the Titus-Bode law.

Since only one solar system is known, we cannot be certain that the above six properties, must all be included in a satisfactory theory or that one or more resulted from an unlikely accident. However, a number of stars are known to have companions with masses comparable to that of Jupiter, and although only perturbations in the path of the stars can be observed it is probable that other planetary systems do exist.

During the history of speculations regarding the origin of the planets, two general lines of thought developed. One held that a catastrophe occurred, involving an encounter between the sun and another object or pair of objects. Such *catastrophic* theories were initiated by Buffon in 1745 (for various early references see Jastrow and Cameron, 1963, or Williams and Cremin, 1968) who proposed a collision with a comet. Colliding stars and colliding nebulae were later considered and Jeans showed that a close approach would serve in place of a collision. Such theories became very popular when Jeffreys showed in 1918 that the planets could not possibly be formed from a condensing gaseous mass, whatever density distribution was allowed. The word "possibly" used here does not include electrodynamic effects.

Many variations of the collision hypothesis were considered, one even involving a solar companion which became a nova. This work continued until about 1946 but met with ever-increasing difficulties. Of the various hypotheses advanced, only that of Dauvillier in 1942 requires mention here. In order to stabilize a tidal filament drawn out from the sun, he suggested that it was ionized and invoked electromagnetic effects.

The second line of thought might be termed the *condensation* hypothesis which began with the work of Descartes, Kant and Laplace. These theories take two forms: the first starts with a fully developed sun surrounded by a gaseous envelope, while the second starts with a nebula which provides not only the planets but the sun itself. Starting with a gas cloud around a fully developed sun, it is possible, without invoking electrodynamic forces, to explain many features of the planetary system such as the mass distribution and regularity or orbits. In spite of difficulties mentioned below Berlage (1948) has continued to develop the theory in a long series of papers which explain the Titus-Bode law and which even included electromagnetic forces in a rather elementary way. Another theory which starts with a fully developed sun is that of Alfvén (1954), whose model is developed almost entirely around electromagnetic forces and so requires some discussion here.

Neutral atoms in the gaseous envelope fall towards the sun and, depending on their ionization potentials, different elements are stopped at different distances by being ionized and then trapped in the solar dipole magnetic field. Alfvén distinguishes clouds consisting mainly of helium, hydrogen, carbon and silicon plus iron. Impurities in the helium cloud condense into Mars and the moon (later captured), and other clouds provide other groups of planets and later the satellite systems are formed in the same way. This theory, like Berlage's, suffers from the difficulty that it leaves the formation of the sun unexplained. Nevertheless, it is likely that features of Alfvén's model and of Berlage's model should be retained even if the models are replaced by the more attractive nebular hypothesis.

According to the *nebular* hypothesis, the sun and planets develop together from a self-gravitating nebula made of the gas and dust of interstellar space. An interesting corollary of this type of theory is that planetary systems are common and this has led to many discussions of the possibility of life outside our solar system (see, for example, Sullivan, 1964). As the protosun contracts it must spin faster in order to conserve angular momentum, and must shed mass in the equatorial plane. It is presumed that the planets are formed from the material of the disk (Kuiper, 1951). This nebular condensation theory fits into the context of modern ideas on star formation; it naturally predicts circularity and coplanarity of orbits, and seems much the most promising of all theories.

The main problem met is the distribution of *angular momentum*. The sun should have preserved most of its original angular momentum as it contracted and have the greater share, whereas it has only a tiny fraction. This difficulty led to the virtual abandonment of the condensation hypothesis for 150 years, but it is now overcome by taking into account the effects of a magnetic field which must be frozen into the protosun and also into the surrounding gas cloud. The possible significance of a magnetic field was first suggested by Birkeland in 1912 and has been discussed by Alfvén (1954), ter Haar (1949) and Hoyle (1960). A ring of material is ejected from the contracting sun, the two remaining connected by a magnetic field. A torque develops between the disk and the faster rotating sun and Hoyle finds that a field of strength about 1 gauss is sufficient to remove most of the angular momentum from the sun. The sun then continues to collapse while the disk moves outwards and rotates less slowly. An alternative approach to the angular momentum problem has been made by McCrea (for references, see Williams and Cremin, 1968), who envisaged the contraction of a cloud of mass sufficient to form several hundred stars. Instead of assuming homogeneity, the cloud is broken into many cloudlets with random motions of about 1 km/sec. Some cloudlets coalesce to form a star of low angular momentum, while others are captured by its gravitational field and become planets. The magnetic field answer to the angular momentum problem is discussed further in Section 11.2 in connection with stars in general. Seen in this broader perspective, and linked with other problems of stellar evolution and activity, it seems almost certain that this is the correct answer.

A second difficulty met in the nebular condensation theory is that a substantial fraction of the nebular mass should have remained in the disk, whereas the planets have only about 0.1 % of the total. It seems likely that some variant of the solar-wind phenomenon (Section 6.1) may account for this discrepancy. In any case this problem is related to that of the condensation of planets and satellites from the original gas of the disk, and here we must lean heavily on the findings of the cosmochemists.

Cosmochemists Urey, Anders and others (for references, see Williams and Cremin, 1968) have proceeded from the fact that some elements and compounds found on the earth and in meteorites are not compatible with the earth having condensed directly from the hot disk gas. The separation of the elements of the earth from the main gas component, hydrogen, must have taken place at low temperature, otherwise the hydrogen would have combined chemically. The picture seems to be that dust forms in the disk, shielding it from the solar radiation and allowing rapid cooling. From the cold gas chondrules, metal grains and other solid particles then formed. These solid particles agglomerated to form larger bodies, the planets and the satellites, whose compositions would depend on the degree of separation of hydrogen and heavier elements which would depend in turn on the distance from the sun.

Cosmochemistry is only of incidental interest here, but it is a fascinating study and may eventually provide the many further clues necessary to determine the complex history of the solar system.

10.2   Jupiter and its Magnetosphere

Jupiter is the largest planet, having a mean radius of about 7×10^4 km, a rotation period of only about 10 hours and a surface escape velocity of 61 km/sec. Its orbit is at 5.2 AU and it has the largest number (twelve) of natural satellites. Its surface is not visible, but it is thought that the bulk of the surface layers are made up of solid hydrogen and solid helium. Bands of clouds rotate at different speeds but the rotation of the solid body of Jupiter is thought to be given by the System III longitude, which is deduced from observed variations in radio emission; these are controlled by the rotation of a planetary magnetic field.

Following the detection of strong, intermittent radio signals from Jupiter by Burke and Franklin (1955), numerous observational (radio and optical) and theoretical studies of the planet have been made and reviewed by Ellis (1965), Warwick (1967) and Drake (1967). In addition to the expected thermal radio emission, there is a broad spectrum of decimeter and decameter radiation. The former is due to synchrotron radiation by electrons of energies in the range 10–100 MeV moving in a magnetic field of strength about 1 gauss. These electrons constitute a Van Allen type of radiation belt, except that the flux is about 10^3 times larger; also the field is stronger, having a strength estimated at 15–20 gauss near the poles. A feature of the decametric emission of notable significance is that it is under the control of the satellite Io, so that the intensity varies according to the relative positions in longitude of both Io and the north magnetic pole (Section 10.3).

The radio emissions, while of considerable interest themselves, are more important in indicating underlying electrodynamic effects which must be present to provide them. First there is the problem of the origin of Jupiter's radiation belt; at a distance from the sun about five times that of the earth, the solar wind is much weaker and less likely to provide a strong radiation belt. Alternative energy sources are the rotation of Jupiter itself and the motion of Io through its magnetosphere. A second problem is the origin of the shock wave or other phenomenon which must be invoked to explain the decameter radio bursts, just as similar effects provide solar radio bursts.

A model of the ionosphere, based on heating and ionization by solar radiation, has been given by Gross and Rasool (1964). They find a temperature of only about 150 K, a scale height less than 100 km and a maximum electron density less than $10^6/cm^3$. The extent of such an ionosphere in hydrostatic equilibrium is small and there would be a negligible plasmasphere at the orbit of Io, which is at about 6 planetary radii (R_j). However, as in the case of the earth's plasmasphere, there may be other factors involved; these are discussed below and in Section 10.3.

One effect which may well be important is Jupiter's rapid rotation. A cloud of plasma trapped in its magnetosphere must rotate with the planet and so experience a large centrifugal force which is in the opposite direction to the gravitational force. If the plasma moves outwards it gains gravitational potential energy but loses pseudo-potential energy associated with rotation. Melrose (1967) has derived an expression for Γ, the total potential energy per unit mass of plasma which rotates with the planet. His results are plotted in Figure 10.1, where Γ is in units of work done on unit mass against gravity in moving (without rotation) from Jupiter's surface to infinity. It is assumed that the plasma moves along a particular magnetic field line of a dipole field; the line cuts the planet at latitude $60°$ and the equatorial plane at a distance $r = 4 R_j$. The scale S represents the distance moved along the field line in units of R_j. The value of Γ is dominated by the gravitational force near the surface of the planet and by the centrifugal force at $4 R_j$. The changeover occurs near $r = 2.5 R_j$ where there are two potential peaks, as shown. Ionospheric particles which have enough kinetic energy to reach these peaks will then "fall outwards" into the

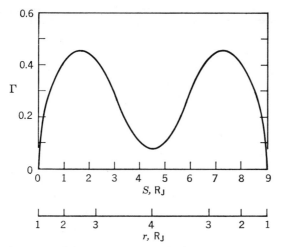

Fig. 10.1. The sum Γ of the gravitational potential energy and the pseudopotential energy of plasma rotating with Jupiter's magnetosphere; the unit is the energy needed to move unit mass from the surface of the planet to infinity (without rotation). The distances S and r are in Jovian radii, measured from the surface of the planet. The former is along a magnetic field line which cuts the surface at 60° latitude, the latter is radially.

potential trough. Plasma trapped in this trough and rotating with the planet will provide some interesting effects, discussed below.

Plasma rotating with a planet experiences a gravitational force towards the center of the planet, a centrifugal force away from the axis of rotation and an electromagnetic force $\mathbf{j} \times \mathbf{B}$ or jB perpendicular to the magnetic field $\mathbf{B}$ and current $\mathbf{j}$. In addition there will be a pressure gradient which we neglect, so that a balance of forces perpendicular to $\mathbf{B}$ gives (Angerami and Thomas, 1964).

$$jB = \rho \, \frac{g}{r^2} \, (1 \, + \, 4 \cot^2 \theta)^{-1/2}$$

$$- \rho\omega^2 r R_j \sin \theta \left\{ 1 \, - \, \frac{9 \cos^2 \theta}{1 \, + \, 4 \cot^2 \theta} \right\} \tag{10.2}$$

where ρ is the plasma density, g is the surface gravitational force, θ is the co-latitude, r is the distance in planetary radii R_j, and ω is the angular velocity. The total current $\int j(\theta, r) \, r d\theta \, dr$ is a ring current, similar to that around the earth but with a different origin. It reduces magnetic flux near the planet and increases it further away, thus inflating and expanding the magnetosphere. Near the equatorial plane and far above the level where gravitational force balances centrifugal force, the Lorentz force balances the latter so that

$$j \sim \rho\omega^2 r^4 R_j B_o^{-1} \qquad (10.3)$$

where B_o is the surface field.

A number of interesting effects may result from the presence of plasma in Jupiter's magnetosphere (Piddington, 1967). First, an unstable situation develops when the field gradient caused by the above ring current equals or exceeds the normal gradient of a dipole field. These gradients are given by the equations $(R = rR_j)$

$$\left|\frac{\partial B}{\partial R}\right| \sim \text{curl } \mathbf{B} = 4\pi \mathbf{j} \qquad \left|\frac{\partial B}{\partial R}\right| = \frac{3B_o R_j^3}{R^4} \qquad (10.4)$$

The instability condition is then found from equations 10.3 and 10.4

$$\rho \gtrsim 3B_o^2 \, (4\pi \, \omega^2 r^8 R_j^2)^{-1} \qquad (10.5)$$

or written in terms of the Alfvén velocity $V_A = B_o(4\pi\rho r^6)^{-1/2}$ and rotational velocity V_R,

$$V_R > \sqrt{3}V_A \qquad (10.6)$$

For example, the maximum proton density consistent with a stable corotating magnetosphere at, say, 30 R_j is about $20/\text{cm}^3$; at 60 R_j it is about $0.1/\text{cm}^3$. Thus any significant amount of plasma which finds its way to the outer magnetosphere will tend to disrupt that part of the magnetosphere.

Since Jupiter's magnetosphere is insulated from the body of the planet, magnetic force tubes may, as in the case of the earth, suffer interchange motions (Section 4.3). Melrose (1967) assumed that the magnetosphere would be stable against such motions and developed a model accordingly. However, there appears no reason why interchange motions should not occur and we conclude that if there is appreciable plasma above a few R_j, then interchange motions will occur. More heavily laden magnetic force tubes will move outwards and be replaced by others; the outward moving tubes must eventually be disrupted and may form a "wrapped around tail" as seen below.

There are other ways that forced interchange motions may be set up in Jupiter's magnetosphere, and these are of interest as possible factors in the development of the radiation belt. First, as in the case of the earth's radiation belt, the belt particles themselves will tend to force interchange motions (Section 9.2). These motions will tend to reduce the flux of trapped particles and so cannot be invoked to explain the origin of the belt. Second, there may be substantial winds in Jupiter's atmosphere which extend up into the ionosphere and create a dynamo electric field as in the case of the earth's ionosphere (Section 3.5). This electric field will project into the magnetosphere (along the field lines) and will create a pattern of plasma Hall drift which is indistinguishable from a pattern of interchange motions of field lines and plasma together. There is evidence of substantial atmospheric motions on Jupiter and these seem to offer distinct possibilities in the problem of the radiation belt. Protons entering the magnetosphere at its outer boundary may acquire 10 keV of kinetic energy as a result of wind and rotational velocities. Dynamo electric fields may then convect them inwards, and at the same time add to their energies.

In order to develop this model further we must consider the possible extent of Jupiter's magnetosphere, which will be limited by the solar wind. If the wind velocity at 5 AU has only fallen to, say, 200 km/sec, then the density will be about $0.3/cm^3$ and for a temperature of $10^5 K$ the magnetic field strength needed to balance the gas pressure is about $1 \gamma (10^{-5}$ gauss). On the solar side, the field must also balance the wind

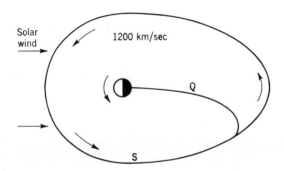

Fig. 10.2. An equatorial section of Jupiter's magnetosphere extending 70 R_J towards the sun and 100 R_J away from the sun. A magnetic field line Q (really a loop from pole to pole) is bent from the meridian plane by the surface frictional effect or by interchange motions.

momentum and the required field is a few γ. Thus, apart from centrifugal forces exerted by magnetospheric plasma, which tend to expand the magnetosphere further, it should extend about 100 R_j away from the sun and about 70 R_j towards the sun. This crude model is shown in Figure 10.2 in equatorial section, with a magnetic field line Q bent out of its meridian plane. If the outer parts of the magnetosphere corotate with the planet, then surface speeds in excess of 1200 km/sec will be experienced, and frictional interaction with the interplanetary plasma is likely, in the same manner that interaction occurs between the earth's magnetosphere and the passing solar wind (Section 7.3).

If, on this model, we now superimpose forced interchange motions, the effects are likely to be drastic. A magnetic force tube moving outwards in the equatorial plane (and to higher latitudes near the poles) may be unable to impart kinetic energy to its plasma corresponding to the velocity of 1000 km/sec and the tube will be wrapped around as shown in Figure 10.3. The whole of the section of the magnetosphere

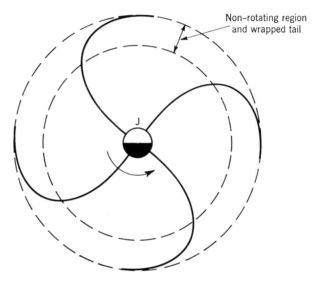

Fig. 10.3. A sketch, in the equatorial plane, of the corotating portion (inside the dashed circle) and non-rotating portion of Jupiter's magnetosphere. In the latter, wrapped around magnetic field lines and a spiral magnetotail are developed.

between the dashed lines may be slowed or stopped, thereby creating shear within the magnetosphere. This, together with the motions of the feet of the field lines through the polar ionosphere, may lead to an active magnetosphere with sufficient plasma density to maintain the activity. As seen in the following section, the satellite Io may contribute to the magnetospheric activity by its powerful electromagnetic interaction. It is also possible that Io has a tenuous atmosphere, as suggested by Binder and Cruikshank (1964).

These studies of Jupiter's magnetosphere and of Io suggest the likelihood of ample rewards for spacecraft measurements of particles and fields near this planet.

10.3   Io and Decametric Radio Emission

Jupiter's satellite Io has attracted much attention for its remarkable control of Jovian decametric radio emission (Bigg, 1964). A number of rather complicated source models have been proposed, but no general agreement has been reached and the interested reader is referred to Ellis (1965), Zheleznyakov (1965), Warwick (1967) and Drake (1967).

Io's orbit is about 6 R_j, deep within Jupiter's magnetosphere and near the outer boundary of the radiation belt. Io and the magnetosphere both rotate eastward, the latter much faster so that Io has a relative westward motion of about 56 km/sec. It seems very likely that a disturbance caused by this motion is responsible for the decametric radio emission, and several possible forms have been discussed. The severity of the disturbance will depend on the plasma density and the magnetic field strength and their combination which gives the Alfvén velocity. If this were less than 56 km/sec, then a shock would form to provide a large disturbance. It seems likely, however, that the Alfvén velocity is far higher than Io's relative velocity and that no shock exists. It is also most unlikely that the plasma density near Io is high enough (about $10^7/cm^3$) to allow the stimulation of plasmas pace-charge waves (Gledhill, 1967). Such an enormous density would be highly unstable to interchange motions and would soon be lost, but in any case its presence should be observable optically as a ring.

More recently the question of Io's disturbance has been discussed by Piddington and Drake (1968) who show that it may depend mainly on Io's electrical conductivity. If this is small, then the satellite will create only a minor disturbance, as does the poorly conducting moon in the solar wind. Irrespective of Io's conductivity, long immersion in the steady magnetospheric field must ensure its magnetization. In fact the magnetic flux through the satellite caused by the field $B \sim 0.05$ gauss (Figure 10.4) will be about 4×10^{15} gauss cm^2, which is about 10% of the flux into a section of the earth's magnetic tail. It would seem that the only way to avoid such permanent magnetization is by the rotation of Io about an axis perpendicular to the external magnetic field. If we reject this possibility, then Io's motion (V) relative to the magnetosphere may have drastic results.

The nature of the disturbance created by Io depends principally on the ratio V/V_A, where V_A is the hydromagnetic velocity. Since V_A probably lies in the range 10^3–10^5 km/sec, this ratio is probably small, and the disturbance has the form shown in Figure 10.4, where the field at the left is compressed and so is diffusing into Io, while that at the right is diffusing out from Io. The bundle of field lines frozen into Io

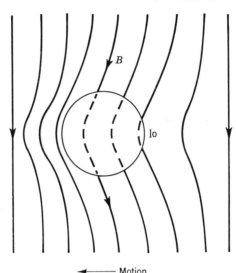

Motion

Fig. 10.4. The passage of the satellite Io through Jupiter's magnetosphere. The dashed field lines are frozen into Io. and perform interchange motions.

is referred to as Io's force tube (IFT), and this is carried along by Io to provide a forced interchange motion. In Figure 10.5 IFT is shown in relation to the planet; the interchange motion is simply a rotation of the force tube about Jupiter's rotational axis NS. In effect Io is exerting a mechanical force on the magnetic tube outwards from the plane of the paper. The force tube is slightly deformed, having its lines of force bent outwards towards Io; the corresponding current system is shown by the arrows marked j. The resulting magnetospheric disturbance is caused by the passage of the whole of IFT rather than by Io alone. If this disturbance is in proportion to the size of the object causing it, then an enhancement by a factor of about 200 is expected. Particles accelerated in this disturbance may contribute to Jupiter's radiation belt.

The two feet of IFT are dragged through the ionosphere along parallels of latitude near $\pm 65°$. They have dimensions about 150 km and speed about 5 km/sec, and may cause substantial heating and a westward travelling shock. With the model ionosphere of Gross and Rasool (1964) a shock is unlikely because of the low ionization density. However, it is possible that the ionization density is very much greater than their estimate, particularly near IFT. Precipitation of fast particles from Jupiter's radiation belt may be substantial and may be greatly enhanced by the IFT disturbance. Such precipitation might increase ionization density to values far above that caused by solar radiation,

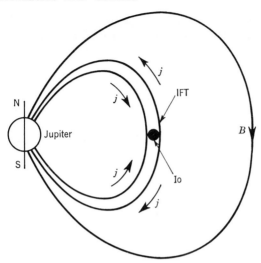

Fig. 10.5. Io and its magnetic force tube move by the complete rotation of both about Jupiter's axis of rotation as an interchange motion outwards from the plane of the paper. In addition the force tube is bent away from the original meridian plane in the manner shown for field line Q of Figure 10.2.

just as in the case of the terrestrial aurora. If this is the case, then an ionospheric shock should form and, by analogy with shock waves in the solar atmosphere, radio bursts may emanate from these fronts in both the forward and reverse directions. According to this model, Io controlled bursts would maximize when Io was approximately at $\pm 90°$ longitude from the Jupiter–earth line; the emission regions would also be much smaller than 1 R_j. The observational data and analysis of Io control (Duncan, 1966) and source size (Dulk *et al.*, 1967) are consistent with these conclusions.

An IFT model, similar to that above, has also been developed by Goldreich and Lynden-Bell (1969). They determine the required Io electrical conductivity σ as follows. The current j (Figure 10.5) flows through Io and through the ionosphere and if Io is to have a maximum effect, then the voltage drop across Io must be much smaller than that in the ionosphere. Taking account of the geometry of the system, the requirement is

$$\sigma \gg \Sigma/L \tag{10.7}$$

where Σ is the integrated (vertically) ionospheric conductivity and L is the dimension of Io. For the Gross and Rasool model ionosphere $\sigma \gg 10^{-17}$ emu, but as we have seen, this may need upward revision.

Goldreich and Lynden-Bell suggest that the radio emission is Doppler-shifted cyclotron radiation from bunches of electrons moving up and down IFT.

10.4 Venus, Mars and Saturn

Venus

The 1962 Venus mission by spacecraft Mariner 2 (U.S.) had an encounter orbit too far from the planet to reveal anything of its plasma environment. However, in 1967 two space capsules arrived near the planet within a period of two days. Venus 4 (U.S.S.R.) arrived first and plunged into the atmosphere, where it recorded pressures rising to some twenty times our sea-level atmospheric pressure and temperatures up to about 270°C. Since transmission then stopped, it was assumed that these measurements gave surface values. However, more recent measurements of the planet's radius by radar suggest that the transmission may have stopped when the spacecraft was still about 24 km above the surface. If so, then the surface pressure might be as high as 100 atmospheres and the temperature also much higher.

Mariner 5 (U.S.) made an approach to within about 10,000 km of the center of Venus and measured magnetic fields, plasma, fast particles and other quantities. The results of the plasma and magnetic field experiments have been summarized by Bridge *et al.* (1967). Abrupt changes in the magnetic field strength and amplitude of fluctuations, and also in the plasma properties, provide clear evidence for the presence of a bow shock around the planet. This is similar in some ways to that around the earth, but is much smaller and appears to result from interaction between the interplanetary magnetic field and the iono-sphere or plasmasphere of Venus. No planetary field could be detected and an upper limit to the magnetic dipole moment of Venus is estimated as 2×10^{-3} that of the earth, corresponding to a surface magnetic field of 1.4×10^{-4} gauss at the poles. The energetic particle experiment (Van Allen *et al.*, 1968) showed that the environment of Venus is almost, if not completely, devoid of energetic particles associated with the planet. This means that the magnetofluid dynamical interaction of the solar wind with the planetary plasmasphere is too weak to generate an observable intensity of electrons of energy exceeding 45 keV.

The curious features of the solar wind and interplanetary field interactions with Venus are illustrated schematically in Figure 10.6. A rough approximation of the Mariner 5 trajectory passes through the five points marked 1–5, each corresponding to a notable change in the

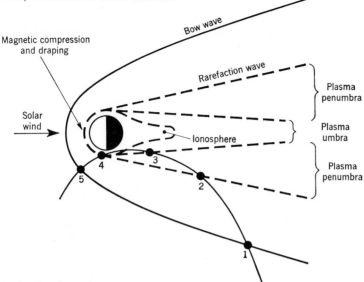

Fig. 10.6. A schematic representation of the interaction between Venus and the solar wind. Magnetic field is compressed in the region between those marked bow wave and rarefaction wave. However, in the region between the spaceprobe positions 4 and 5 the agitated field grew to a maximum value and decreased steadily to the interplanetary value without shock surfaces. The magnetic field is held off by an extensive ionosphere, which also creates a plasma shadow.

field or plasma. At position 1 a shock wave is traversed, the most notable feature being an abrupt increase of magnetic field strength to double the interplanetary value. In this region the bow wave, shown by the full line, has characteristics similar to those of the earth's bow wave. Between positions 1 and 2 the field is fairly steady, decreasing from about 10 γ to about 5 γ. At position 2 the field strength decreases abruptly to a very low and highly variable value indicating a passage through a rarefaction shock. The reason for this shock is the existence of a plasma cavity behind Venus; this is a region of extremely low plasma and magnetic pressure which is invaded from the surrounding region of much higher field and plasma pressure. Near position 3 some plasma is still present but its velocity has fallen to less than half that of the solar wind. This plasma might be considered as part of the Venusian ionosphere or plasmasphere being "blown away" as a result of frictional interaction with the plasma which flows past points 2 and 4 with nearly the full velocity of the solar wind.

At point 4 we might expect to meet the rarefaction shock for a second time, but the nature of the "magnetosheath" between points 4 and 5 is quite different to that between points 1 and 2. The magnetic field direction and strength near both of these points fluctuate violently, but the average strength is approximately that of the interplanetary field. Moving from point 4 to point 5 the field strength rises to about double the interplanetary value and then, still fluctuating rapidly, falls again to equal that value. Since there was only one crossing of this particular region one cannot be certain that the conditions encountered were representative. However, if we accept them as average, then it would seem that the magnetosheath differs greatly from that near the earth. While there was clear evidence of shock surfaces at positions 1 and 2, the field strength and plasma density and velocity at positions 4 and 5 show no abrupt changes. The intermittent appearance of shocks might be inferred from large and rapid changes in the direction of the magnetic field, but this does not appear certain, and in any case the shock does not seem to be a permanent feature of the magnetosheath. The theoretical justification of a bow shock is based on the analogy in fluid dynamics, but this analogy is imperfect and it has been argued that a strong (non-shock) hydromagnetic disturbance may occur on occasions (Piddington, 1964).

Measurements of the atmosphere and plasmasphere of Venus were made during the Mariner 5 bypass, using radio occultation techniques. Mariner Stanford Group (1967) used dual-frequency radio transmissions from the earth which reached the spacecraft after passing through Venus' atmosphere and plasmasphere. The signals were stored on tape and later transmitted back to earth where they revealed both daytime and nighttime ionospheres (or plasmaspheres). Above the former a sharp plasmapause exists, marking a transition from appreciable electron density near Venus to the tenuous conditions of the solar wind. This boundary is shown by the dashed line on the day side of Venus in Figure 10.6. The nighttime ionosphere has a peak electron concentration lower by about 100, but it then extends with gradually diminishing density to beyond 10,000 km as shown in Figure 10.6. Other radio occultation measurements were made by Kliore *et al.* (1967), who used a transmitter on the spacecraft to send signals through the atmosphere and back to earth. The results showed an ionosphere with at least two maxima, and also provided profiles of temperature and density in the neutral atmosphere.

Mars

The 1965 Mars mission by Mariner 4 (U.S.) yielded results generally similar to those found by the later Venus mission, but much less definitive. The approach was only to within about 13,000 km and no magnetic (Smith *et al.*, 1965) or fast particle effects were observed.

However, an atmosphere and an ionosphere exist, and it is probable that a stagnation field builds up (as in Figure 10.6) and deflects subsequent lines and particles. References to the Mars particle results are given in the above Venus references.

A feature of the interaction between the magnetized solar wind and the planetary atmosphere which we have not mentioned is that the atmosphere may be ionized by collisional effects. Some of the various boundary effects discussed in Section 7.3 may operate here, although in the case of Venus and Mars the plasmaspheres are non-magnetic and so there can be no magnetic neutral sheet. On the other hand, Rayleigh-Taylor and two-stream instabilities may be possible and the problem is likely to attract many theorists. Dessler (1967) has discussed the ionizing effects of a standing bow-shock wave at 1.4 times the radius of Mars on the day side.

Saturn

Saturn rotates at almost the same rate as Jupiter and is only a little smaller, and so it may have a magnetic field. If this is so, then the interaction of the field with the planet's rings could provide interesting effects, some of which have been discussed by Zheleznyakov (1964) and Zlotnik (1967). These writers were mainly concerned with the possibility of distorting a dipole field so as to confine trapped energetic particles to regions of high latitude. Another possible effect of interest is the creation of ionospheric heating and disturbances by the relative motions of the planet and its rings. The two plasma regions would be connected magnetically in a manner similar to that of Io and Jupiter (Section 10.3).

10.5 Lunar Electrodynamics

Presumably on the basis of its relative proximity, numerous statistical studies have been made of possible lunar influence on the geomagnetic field. If substantiated, these might indicate electrodynamic interaction and so would be of interest here. However, in a review Schneider (1967) concluded that the early claims of an astonishingly large effect are without foundation, and there remains but marginal evidence of a small effect near full moon. The observational data discussed below would seem to substantiate this conclusion.

Prior to the direct measurement of magnetic fields and particles in the vicinity of the moon, it was conjectured that a fluctuating magnetic field region in a sheath layer would develop on the solar side. The model resembled that of Figure 10.6, except that the field was blocked by a solid, electrically conducting body rather than by an ionized atmosphere. However, direct measurements by lunar-orbiting Explorer 35 revealed no such layer and have led to an entirely different picture. These measurements are so much more detailed and accurate than earlier ones that the latter are now of only historic interest.

Magnetic field measurements (Ness *et al.*, 1967) reveal no magnetic field change on the solar side of the moon, and only small perturbations on the anti-solar side (Colburn *et al.*, 1967). Related plasma measurements show a shadow region behind the moon in which no solar plasma flow could be detected. These results may only be interpreted in terms of a moon of low electrical conductivity, at least in an outer shell, through which the magnetic field lines may move freely, while the plasma flow is cut off. The upper limit of the conductivity given by the diffusion equation 2.12 is about 10^{-16} emu. We now have three distinct types of interaction between the solar wind and various objects. The earth is surrounded by an extensive magnetosphere which holds off the wind and creates a magnetosheath; Jupiter may provide a somewhat similar interaction complicated by rapid rotation. In the case of Mars and Venus the magnetosheath is held off by an electrically conducting plasmasphere, also able to withstand the wind. In the case of the moon, the field lines pass right through the solid body. The upper limit of lunar conductivity determined from this model is not inconsistent with measurements of the conductivities of materials commonly found in the earth's crust by Hamilton (1965). Alternatively, only the surface shell may be a poor conductor as it is a poor thermal conductor because it comprises small pointed grains of material (Jaeger and Harper, 1950).

The removal of plasma behind the moon creates a perturbation in the interplanetary magnetic field which has attracted considerable theoretical interest. A qualitative explanation is simple, at least in the cases where the magnetic field is parallel to or perpendicular to the wind velocity. Immediately behind the moon there is a plasma vacuum or umbra, and plasma beyond the boundary of this shadow region begins to move in with thermal velocity to fill the region. With the magnetic field parallel to the wind velocity the "implosion" becomes a hydromagnetic expansion wave which reduces the magnetic pressure and plasma pressure near the boundary, and increases the magnetic pressure, but not the plasma pressure, in the shadow region. Hence we have a conical umbra of increased field strength and no plasma, and a penumbra of decreased field strength and reduced plasma pressure. The situation is somewhat similar to that behind Venus, shown in Figure 10.6.

A quantitative evaluation of the moon's electromagnetic field and plasma distribution is difficult, and resort has been made to a number of approximate treatments. One such treatment is based on the assumption that the solar plasma behavior is described by a fluid flow model. As we have seen in Section 2.4, such a treatment of a collisionless plasma is valid only in the case of a two-dimensional model, that is when there are no gradients in the direction of the field. This treatment has led some workers to predict a shock wave behind the moon, while other treatments and observational results provide no evidence of a shock. It is interesting to recall that the same treatment led to the prediction of shocks on the solar side of the earth and of Venus, while other considerations indicated that it might be possible for shocks to be avoided, and observations seem to indicate their absence at times. An approximate fluid-flow method has been used by Johnson and Midgley (1968), who show that a permanent shock might form ahead of the moon if its surface conductivity is high, a temporary shock might form if it has a very large conducting core, and no shock will form if there is a thick surface layer of poor conductivity.

An alternative treatment, which might be expected to yield more generally accurate results, is that of Whang and of Ness et al. (1968, where earlier references are given). This is based on a guiding-center approximation, the plasma being treated as a collisionless gas with one-dimensional thermal motions of the protons only along magnetic field lines. Three perturbation fields are determined: the field due to particle gyration around the unperturbed field, the field due to particle gradient drift, and the field due to particle curvature drift. The perturbed field and particle motions may then be ascertained. A long, non-axially symmetric wake region develops in which a plane of symmetry exists, defined by the local solar wind velocity and interplanetary magnetic field direction. There are no shock surfaces, but rather a plasma umbra surrounded by a penumbra in which the density increases steadily from zero to that of the unperturbed plasma.

10.6 Comets and the Solar Wind

It is believed that comets have solid nuclei of diameter about 10–20 km consisting of dust and molecules of C, N, O and H. The molecules form a frozen block, into which the dust particles are frozen (Whipple, 1951). If a comet approaches the sun to within a few astronomical units, it develops a coma of diameter roughly 10^5 km comprising non-ionized gas molecules CN, C_2, NH and others. In addition, it grows one or two tails of which there are three types. Type-I tail consist of CO^+,

N_2^+, CO_2^+ and other ions, Type-II and Type-III tails are composed mainly of dust particles and non-ionized gas. Type-I tails have a diameter about 10^5 km and lengths up to 10^7 km, the dust tails are generally shorter.

In Chapter 6 we mentioned that comet tails provided some of the earliest evidence of the existence of a permanent solar wind. The evidence is that Type-I tails are straight and point continuously in a direction almost directly away from the sun as the comet rounds the sun. Type-II and Type-III tails are curved and trail behind the comet; the acceleration of their material can be explained by the pressure of sunlight. The acceleration of Type-I tails is 10^2 or 10^3 times larger and requires an alternative explanation; furthermore it shows some correlation with geomagnetic activity (Biermann, 1951). It would seem that these effects might be explained fairly simply in terms of an interaction between the solar wind and ionic comet tails. However, while this is undoubtedly the correct explanation, the nature of the interaction is most complicated.

The first type of interaction to consider is simply by means of the mass flow of the solar wind, which might be taken as 10^9 protons/cm²/sec. Over the cross sectional area of the coma or tail (diameter 10^5 km) this amounts to about 10^{29} protons/sec. The loss from an average comet is roughly 10^{30} molecules/sec and if the molecular weight is taken as 20, then this mass loss is two orders of magnitude greater than solar wind flow. It would seem that the interaction with the solar wind must occur over an area much greater than the cross sectional area of the coma. The problem here is somewhat similar to that discussed in Chapter 9, where the energy requirement of the geomagnetic tail was seen to indicate a solar wind interaction over an area much greater than that offered by the corotating magnetosphere.

This was not the only problem, however, because it is not clear how interaction occurs between the coma and that part of the solar wind which does pass through the coma. Collisions between solar protons and neutral molecules are unimportant because of the small collision cross section; the same applies for collisions between solar protons and cometary ions. It turns out that it is the solar electrons which provide important interaction. These collide with the cometary ions and give up a little momentum as they are stopped. The protons continue on and so an electric space-charge field develops which immediately accelerates the electrons so that they suffer more collisions with cometary ions. Another interaction, which is important in ionizing the cometary molecules, is that of charge transfer, whereby a solar proton passing a neutral CO or N_2 molecule picks up an electron and leaves an ionized molecule. The cross section for this effect is large (about 3×10^{-15} cm^2) and it is important, but alone is an inadequate source of ionization.

An important step in explaining the solar wind comet interaction was the suggestion by Alfvén (1957) of a transverse magnetic field in the solar plasma. This idea was developed by Harwit and Hoyle (1962), who envisaged field and solar particles entering the comet coma of un-ionized gas and creating ionized molecules by charge transfer. The new ions are attached to the field lines and exert a drag, so that lines drape around the comet head in the way they drape around the plasma-sphere of Venus. Solar wind momentum is then transferred through the draped field over an area greater than the coma cross section. This is a considerable step towards explaining the momentum requirement, but as this relatively simple model was further developed it was found to be only marginally adequate.

A further step was made by Axford (1964, where earlier references are given), who suggested that upstream from the comet head the hypersonic solar wind will undergo a shock transition similar to the bow shock upstream of the terrestrial magnetosphere (Figure 6.4). As we have seen, the analysis of this situation using the fluid-flow approximation may lead to serious errors because of the free motions of particles along the magnetic field lines. However, the development of bow shocks near the earth and near Venus provide strong evidence for their appearance near comets. They would help explain several cometary features, the first being the appearance of a roughly parabolic envelope around the comet head; this is the theoretical fluid-flow configuration. The partial thermalization of the solar wind as it passes through the bow shock must result in its loss of momentum over an area much greater than that of the coma, so that the momentum discrepancy may be removed. It is also known that downstream of the earth's bow shock, considerable fluxes of electrons of energies above 1 keV appear. If such electrons also appear behind a comet shock they are likely to have some important effects. First, they will contribute to the ionization of the molecular ions and so compensate for the apparent inadequacy of photo-ionization and of charge-transfer ionization. Second, they may be important in the microscopic process of momentum transfer from the solar wind to the cometary ions. They will collide more frequently and more violently than would the original solar wind electrons (energy about 2 eV), and so may drive the molecular ions along the magnetic field lines to form the tail and the ray structure observed in some tails.

Clearly the various ionization and magnetohydrodynamic processes occurring in and around a comet are complex. A number of models have been investigated quantitatively, usually in the fluid-flow approximation. These neglect electrodynamic effects except in recognizing that they are responsible for the fluid properties of the collisionless gas. The interested reader is referred to Biermann *et al.* (1967), who gives earlier references, and to Ioffe (1968).

An understanding of the interaction of the solar wind with comet tails is not only important in connection with the comets themselves, but also in the study of the solar wind in regions not yet accessible to spacecraft. In this respect Brandt *et al.* (1966) have measured the orientations of a large number of tails and shown that statistically the tail is directed nearly in the plane of the orbit and in a direction measured by the combined velocity vectors of the solar wind and the comet. There is a sharp velocity minimum to the solar wind of 150 ± 50 km/sec, a mean velocity of 500 km/sec, and a correlation between the direction of tails and geomagnetic activity. There is not yet agreement as to correlations between tail directions and solar activity or tail directions and latitude. Furthermore, some planets have tails directed in advance of the radius vector from the sun and others show structural peculiarities which are difficult to explain on the above picture (Öpik, 1964).

There appears to be scope for much further work in this field, including some proposed observations of artificial comets involving the ejection of gas clouds from spacecraft.

References

Alfvén, H., 1954, *On the Origin of the Solar System*, Oxford University Press, London.

Alfvén, H., 1957, *Tellus* **9**, 92.

Angerami, J. J. and Thomas, J. O., 1964, *J. Geophys. Res.* **69**, 4537.

Axford, W. I., 1964, *Planet. Space Sci.* **12**, 719.

Berlage, H. P., 1948, *Proc. Koninkl. Ned. Akad. Wetenschap.* (*Amsterdam*) **51**, 796 and 965.

Biermann, L., 1951, *Zeit. Astrophys.* **29**, 274.

Biermann, L., Brosowski, B. and Schmidt, H. V., 1967, *Solar Phys.* **1**, 254.

Bigg, E. K., 1964, *Nature* (*London*) **203**, 1008.

Binder, A. B. and Cruikshank, D. P., 1964, *Icarus* **3**, 299.

Brandt, J. C., Belton, M. J. S. and Stephens, M. W., 1966, *Astron. J.* **71**, 157.

Bridge, H. S., Lazarus, A. J., Snyder, C. W., Smith, E. J., Davis, L., Coleman, P. J. and Jones, D. F., 1967, *Science,* **158**, 1669.

Burke, B. F. and Franklin, K. L., 1955, *J. Geophys. Res.* **60**, 213.
Colburn, D. S., Currie, R. G., Mihalov, J. D. and Sonett, C. P., 1967, *Science* **158**, 1040.
Dessler, A. J., 1967, *Atmospheres of Venus and Mars*, J. C. Brandt and M. B. McElroy, eds., Gordon and Breach, London.
Drake, J. F., 1967, *Univ. Iowa Res. Rept.* 67–22.
Dulk, G., Rayhrer, B. and Lawrence, R., 1967, *Astrophys. J. Lett.* **150**, 117.
Duncan, R. A., 1966, *Planet. Space Sci.* **14**, 1291.
Ellis, G. R. A., 1965, *Radio Science* **69D**, 1513.
Gledhill, J. A., 1967, *Nature (London)* **214**, 155.
Goldreich, P. and Lynden-Bell, D., 1969, *Astrophys. J.,* in press.
Gross, S. H. and Rasool, S. I., 1964, *Icarus* **3**, 311.
Hamilton, R. M., 1965, *J. Geophys. Res.* **70**, 5679.
Harwit, M. and Hoyle, F., 1962, *Astrophys. J.* **135**, 875.
Hoyle, F., 1960, *Quart. J. Roy. Astron. Soc.* **1**, 28.
Ioffe, Z. M., 1968, *Soviet Phys.* **11**, 1044.
Jaeger, J. C. and Harper, A. F. A., 1950, *Nature (London)* **166**, 1026.
Jastrow, R. and Cameron, A. G. W., eds., 1963, *Origin of the Solar System*, Academic Press, New York.
Johnson, F. S. and Midgley, J. E., 1968, *J. Geophys. Res.* **73**, 1523.
Kliore, A., Levy, G. S., Cain, D. L., Fjeldbo, G. and Rasool, S. I., 1967, *Science* **158**, 1683.
Kuiper, G. P., 1951, *Astrophysics. A Topical Symposium*, J. A. Hynek, ed., McGraw-Hill, New York.
Lyon, E. F., Bridge, H. S. and Binsack, J. H., 1967, *J. Geophys. Res.* **72**, 6113.
Mariner Stanford Group, 1967, *Science* **158**, 1678.
Mayer, C. H. and Spangler, M., 1967, *I.A.U. Draft Reports, 13th General Assembly*, p. 310.
Melrose, D. B., 1967, *Planet. Space Sci.* **15**, 381.
Ness, N. F., Behannon, K. W., Scearce, C. S. and Caterano, S. C., 1967, *J. Geophys. Res.* **72**, 5769.
Ness, N. F., Behannon, K. W., Taylor, H. E. and Whang, Y. C., 1968, *J. Geophys. Res.* **73**, 3421.
Öpik, E. J., 1964, *Z. Astrophys.* **58**, 192.
Piddington, J. H., 1964, *Space Sci. Rev.* **3**, 724.
Piddington, J. H., 1967, *Univ. Iowa Rept. 67–63*.
Piddington, J. H. and Drake, J. F., 1968, *Nature (London)* **217**, 935.
Schneider, O., 1967, *Space Sci. Rev.* **6**, 655.
Smith, E. J., Davis, L., Coleman, P. J. and Jones D. E. 1965, *Science* **149**, 1241.
Sullivan, W., 1964, *We are not Alone*, McGraw-Hill, New York.
ter Haar, D., 1949, *Astrophys. J.* **110**, 321.
ter Haar, D., 1967, *Ann. Rev. Astron. Astrophys.* **5**, 267.
Van Allen, J. A., Krimigis, S. M., Frank, L. A. and Armstrong, T. P., 1968, *J. Geophys. Res.* **73**, 421.
Warwick, J. W., 1967, *Space Sci. Rev.* **6**, 841.
Whipple, F. L., 1951, *Astrophys. J.* **113**, 464.

Williams, I. P. and Cremin, A. W., 1968, *Quart. J. Roy. Astron. Soc.* **9**, 40.
Zheleznyakov, V. V., 1964, *Soviet Astron.* **8**, 765; *Astronom. Zh.* **41**, 955.
Zheleznyakov, V. V., 1965, *Soviet Astron.* **9**, 617; *Astronom. Zh.* **42**, 798.
Zlotnik, E. Ya., 1967, *Soviet Astron.* **11**, 462; *Astronom. Zh.* **44**, 581.

CHAPTER 11

Stars and the Interstellar Medium

Looking beyond the solar system we see a vast conglomeration of stars making up our galaxy. This stellar system has a diameter of about 30 kpc (kiloparsec, a parsec being 3.1×10^{18} cm or about 3 light years), and average thickness about 200 pc. The total number of stars is about 10^{11}, so that their average distance apart is about 1 pc or 2×10^{5} AU. The sun itself is a very average star in both mass and brightness. Far more massive and bright are some recently formed O and B type stars, while at the other extreme are old, dying or dead stars which have collapsed to planetary dimensions or less.

Between the stars is the interstellar medium comprising *gas, magnetic field, cosmic rays* and *dust*. The interstellar material contributes only a few percent of the total mass of the galaxy, but it plays an essential part in its evolution. During the lifetime of the galaxy, one or two times 10^{10} years, all of the stars have formed from interstellar material and new stars are still being formed. This process also operates in reverse, gas being ejected from stars by various processes. Furthermore, some of the ejected gas has undergone thermonuclear transformation and so the primeval hydrogen–helium gas is enriched with heavier elements.

Pervading the whole system of gas and stars and extending far beyond is a magnetic field which, as we shall see in the following chapter, evolves and strengthens as the galaxy ages. This field is responsible for various magnetohydrodynamic effects of fundamental importance.

(1) It has considerable control over gas motions and the formation of clouds of higher than average density.

(2) It plays an important, perhaps essential, role in the birth of some stars.

(3) It is responsible for some interesting electrodynamic effects within and around stars.

(4) It is responsible for the acceleration of some of the galactic cosmic rays.

(5) It is responsible for numerous radio emissions from a variety of galactic objects; it is also probably responsible for some x-ray and gamma-ray emissions.

The third main constituent of the interstellar medium is the cosmic-ray gas. This was not previously regarded as an object of astronomical interest but has come increasingly to the fore as its significance in radio and x-ray sources, and also in its control of hydromagnetic motions, became clear. The interplanetary medium and some important dynamical processes involving this medium are discussed in Section 11.1.

One such process is formation of gas clouds, some of which continue to contract until they fragment into protostars which then commence their long process of evolution. The theory of *star formation* is discussed very briefly in Section 11.2. For a much more detailed discussion the reader is referred to Spitzer (1969).

After they have formed, stars pose many interesting problems in electrodynamics. These include interaction between the various force fields, including the electromagnetic field, magnetic braking of stars and the observed magnetic variations. These are discussed in Section 11.3.

In addition to gas clouds and stars, many other galactic objects reveal complex electrodynamic effects. Some of these objects are distinguished by their radio emission, others by x-ray emission and in some cases even gamma-ray emission. While these "exotic" forms of electromagnetic radiation may, like most optical emission, have a thermal origin, in many cases the mechanism responsible involves electromagnetic fields. Some of these objects are supernova remnants. One of the objects is the galactic radio corona; another is the galactic central region which has a most complex structure. The galactic neutral hydrogen distribution may be studied by its line emission and supernova remnants by their synchrotron emission. The various emissions and their sources are discussed in Section 11.4.

Of all the discrete sources of non-thermal radiation, that which has puzzled astrophysicists for the longest is the notorious *Crab Nebula*. Even before the development of radio and x-ray astronomy, the optical emission could not be explained by any reasonable thermal model. This object is now the most fully documented of all the sources, radiation having been measured over a spectral range of 10^{12} to 1. The nebula has the added distinction of revealing violent magneto-hydrodynamic activity and probably being a region of magnetic amplification. For these reasons a whole section (11.5) is devoted to this single object.

The chapter is completed with a brief review of our knowledge of those strange radio sources, the *pulsars*.

11.1   The Interstellar Medium

The interestellar gas and dust are important primarily because they are the materials from which stars were made and are still being made. The chemical composition of these materials is changing because of nucleosynthesis in stars followed by ejection of gas by various processes, notably nova and supernova explosions. As we shall see in the following chapter, the magnetic field and cosmic-ray content of the galaxy also change in an evolutionary manner and so the characteristics of the interstellar medium may differ greatly from those of earlier times. We must conclude then, that theories of formation of gas clouds and stars and theories of stellar activity discussed below differ, at least quantitatively, according to the epoch to which they refer. For example, stars formed while the galaxy was young may have lacked substantial magnetic fields and this may have reduced the prevalence of planetary systems (Section 10.1).

Interstellar Gas and Cosmic Rays

The density of the interstellar gas ranges from considerably less than 1 atom/cm^3 to 10–10^3 atoms/cm^3 in the clouds. The gas is mainly hydrogen, about 10% being ionized, mostly in HII clouds which surround early-type stars (Strömgren spheres). There is also a very small proportion of metal atoms, which is important because of the low ionization potential of these elements. This ensures that the gas even in rather cold HI regions is electrically conducting and able to freeze in the magnetic field (a condition which may not have obtained when the galaxy was young). Thus all motions other than of scale less than about 0.1 pc are hydromagnetic motions of the lightly ionized gas (Piddington, 1957, p. 515).

The interstellar cosmic-ray density is generally considered to be approximately constant throughout the galaxy and equal to the value near the earth, after allowing for some swept away by the solar wind. However, as seen below, it is difficult to account for the galactic radio synchrotron emission without increasing at least the electron cosmic-ray density. The question of the distribution of all galactic cosmic rays is discussed in Section 12.5 in connection with the origin of this radiation.

Interstellar Magnetic Field

An interstellar magnetic field was proposed by Alfvén in an attempt to explain the cosmic-radiation density observed at the earth. In 1949 Fermi showed ways in which such a field would not only help contain the cosmic rays, but would also accelerate them. The first direct evidence of such a field came from Hall and Hiltner, who in 1949 demonstrated the polarization of star fields. This is caused by the alignment of the interstellar grains by a magnetic field. The existence of this field has been amply confirmed by radio-astronomical observations of synchrotron radio emission and of Faraday rotation of plane polarized radio waves (for a review, see Gardner and Whiteoak, 1966).

The strength of the field is another matter, and has been a matter of controversy ever since its discovery (for reviews, see van de Hulst, 1967, and Spitzer, 1968). The highest estimates are based on the interpretation of meter wavelength radiation as synchrotron emission by electron cosmic rays of density everywhere equal to that measured at the earth. The required field is 3×10^{-5} gauss, which is an order of magnitude above that found by other methods. The field found by Faraday rotation measurements is a few times 10^{-6} gauss, while a substantially larger value (about 2×10^{-5} gauss) is found in one particular direction by the Zeeman splitting measurement (Verschuur, 1968). As seen below, further evidence of a field of about 3×10^{-6} gauss or less is that a stronger field would cause too much expansion of the galactic disk. More recent estimates, using the radio emission from pulsars (Section 11.6) give somewhat lower values. Perhaps the best estimate of average field strength near the sun is 3×10^{-6} gauss, the direction of the field being more or less perpendicular to the galactic radius. In order to account for the synchrotron emission, either the field strength or the cosmic-ray density or both must increase in the general direction of the galactic center. (see Davies et al., 1968; Mathewson and Nichols, 1968).

Interstellar Medium Dynamics

The dynamics of the interstellar medium are complex and here we can give only the briefest of reviews. As we have seen above, motions within the medium are hydromagnetic motions of the whole gas. When the gas is fully ionized these motions are little damped, but in HI regions the damping is severe because of collisions between ions and neutral atoms. However, the plasma is continually being disturbed in a variety of ways so that new motions are generated as the old decay.

A new (O or B) star emits ionizing ultraviolet radiation which is all absorbed within a sharply bounded region called a Strömgren sphere. Within the sphere the gas is nearly all ionized and at a temperature of about 10^4K. The Strömgren radius around a star of surface temperature 50,000K is 90 $N^{-2/3}$ pc, where N is the proton density; that for a star of surface temperature 11,000K is only 0.5 $N^{-2/3}$ pc. Outside these regions are cold HI regions where the pressure is smaller by factors of order 100, and so the HII regions expand and set the HI gas in motion and compress it to densities which may be sufficient to start star formation.

Other sources of interstellar disturbance are supernovae, whose shells move outwards with velocities of 1–3×10^3 km/sec and compress the interstellar gas and magnetic field. These disturbed regions are distinguished from HII regions by much higher electron temperatures, up to 10^5K, and by a highly filamentary structure. Yet other disturbed regions are planetary nebulae, which result from less violent ejection of gaseous envelopes from various types of star. The physics of interstellar disturbances has been reviewed by Kahn and Dyson (1965), Spitzer (1969) and Pikel'ner (1968).

The interstellar magnetic field is frozen into both the ambient plasma and the cosmic-ray gas, and so the three components are linked together and exercise control over one another. On the one hand cosmic rays may abstract energy from the hydromagnetic waves, as first suggested by Fermi. Some of the disturbances develop into collisionless shock waves and these are particularly effective in accelerating cosmic rays. In other situations the cosmic-ray gas may expand and give up its energy to the ambient plasma. Such a case is found in the Rayleigh-Taylor instability of spiral arms under the disruptive forces of magnetic and cosmic-ray pressures.

Parker (1966, 1967) has investigated the equilibrium of the gas in the galactic disk under the opposing influences of the gravitational force of the stars (towards the galactic equatorial plane) and of gas pressure p, cosmic-ray pressure P and magnetic pressure $B^2/8\pi$ (field parallel to the plane). For equilibrium we have

$$\frac{d}{dz}\left(p + P + \frac{B^2}{8\pi}\right) = -\rho(z)g(z) \tag{11.1}$$

where z is the distance above the plane, ρ is the gas density and g is the gravitational acceleration. Included in p are the mass velocities of the gas clouds, and if we put $P = 0.5 \times 10^{-12}$dyne/cm^2, $B = 5 \times 10^{-6}$ gauss, the gas density needed to give the observed scale height is about 7 atoms/cm^3, which is much larger than the generally accepted estimate of about 1 atom/cm^3. The discrepancy is not serious, however, because the density requirement may be reduced by reducing B, by the presence of unobservable molecular hydrogen, by twisting the field into a spiral, or by the presence of a magnetic field outside the disk. Nevertheless, the analysis provides further evidence against the highest estimates of the magnetic field strength.

The more interesting result of Parker's analysis is that the equilibrium is unstable against transverse hydromagnetic waves with perturbation fields perpendicular to the galactic plane. A small perturbation causes gas to flow along the field into the region of the wave trough; the upper bulge is thus released and expands further while the lower region is further compressed as shown in Figure 11.1. This is the Rayleigh-Taylor

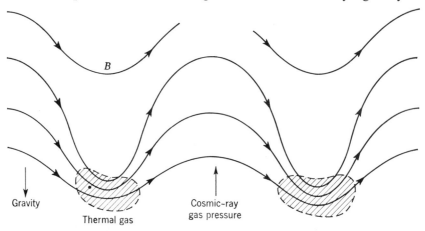

Fig. 11.1. A sketch of the Rayleigh-Taylor instability of the interstellar gas, which is supported against gravitational force by the pressure of the magnetic field B and the cosmic-ray gas.

instability, where a more dense fluid supported against gravity by a less dense fluid, drips downwards through the latter. The onset of the instability depends on γ the usual ratio of specific heats given by $\delta p/p = \gamma(\delta\rho/\rho)$, and also on the proportions of magnetic and cosmic-ray pressures to the total pressure. Even with $\gamma = 4/3$, a value which would normally ensure stability, the above-mentioned conditions lead to some instability. In the galaxy and for the time scales in question ($\gtrsim 10^7$ years) the value of γ is probably closer to unity and the medium is highly unstable. The time scale of the effect is the time of free fall, about 3×10^7 years, which is only a fraction of the rotation period of the galaxy at the sun. The size of the gas clouds is 10–10^3 pc, which is in agreement with observed sizes. Parker (1968) has further extended the theory in later papers.

11.2 Star Formation

In the preceding section it was seen that the gas and dust in our galaxy is irregularly distributed. It is only within the regions of high gas and dust density that high-luminosity stars are seen. Sometimes groups of these stars are seen receding from one another with velocities which indicate that the associations formed not more than a few million years ago, a fact which receives confirmation from the theory of thermonuclear reactions in stars. These three lines of evidence prove conclusively that stars are being formed now in our galaxy and in other spiral systems, in regions of high gas and dust density. These young stars make up Baade's population type I; in addition there are old stars of population type II which were formed as long ago as 5×10^9 years.

Because of their relatively recent birth, in conditions which are observable, it is the young stars whose formation provides the most fruitful study. These stars also provide a time scale for the formation and disappearance of spiral arms which are made up of young stars (Section 12.2).

The gas clouds within which and from which stars are formed are made up principally of neutral hydrogen of density 10–100/cm^3 and temperature 100K. The extent of such clouds is typically 10 pc and the corresponding mass 300–3000 $M_{\odot}$ (solar masses). There are sufficient atoms of low ionization potential to give electron densities about 10^{-4} gas density, which provides sufficient electrical conductivity to partially freeze in the magnetic field of strength perhaps a few times 10^{-6} gauss.

Star formation is thought to occur in the following stages:

(1) Suppose that an HI cloud at temperature 100K is surrounded by HII gas at temperature 10^4K and, of course, much lower density. The pressure p is the same on both sides of the boundary between the two gases and if p increases, the radius R of the cloud decreases and the gas is compressed isothermally. At a particular value of R gravitational forces take control and the cloud collapses. This occurs when the gravitational potential energy exceeds the thermal kinetic energy by a factor of order unity, or more precisely when

$$MR^{-2} > 5(p/G)^{1/2} \qquad (11.2)$$

where M is the cloud mass and G is the gravitational constant.

There are many clouds in the galaxy which appear to satisfy the above requirement of instability. However, the presence of a magnetic field changes the picture radically. It is shown in Chapter 12 that if a magnetized cloud is in equilibrium between magnetic pressure and gravitational inward force, then it remains in equilibrium no matter how much it contracts; it is in a sort of neutral equilibrium and never collapses. Instead of a critical radius, there is a critical mass and for a field of 2×10^{-5} gauss this is 10^7 $M_\odot$ which is far too great. For a field of 2×10^{-6} gauss it is 10^4 $M_\odot$ which is in rough agreement with the observed cloud complexes. Thus we have further evidence for the low value of the interstellar magnetic field.

(2) The second stage is one of collapse and fragmentation into protostars. The process is difficult to analyse because the parameters change over a wide range: the radius decreases from 20 pc to 0.02 pc while the magnetic field increases from 2×10^{-6} gauss to 2 gauss. Also the effects of increasing turbulence and angular velocity are complications. However, the central problem is the magnetic field, which is important in several ways. First, as seen above, the field determines a critical mass which depends also on density, and so the field will determine the scale of fragmentation and the very nature of the stars themselves.

Second, a major problem is met in keeping the magnetic field strength within reasonable limits while increasing the gas density from that of the initial cloud to that of the final star, a factor of 10^{22}–10^{23}. A fully frozen-in field would increase from 2×10^{-6} to about 10^9 gauss, which is prohibitively large. Some field lines may diffuse out of the cloud if the electrical conductivity is reduced sufficiently; this may occur if the cloud becomes optically thick so that no starlight reaches its core and even the metal atoms are neutral. However, this effect appears marginally adequate (Spitzer, 1963), and Mestel (1966) has proposed the alternative of plasma contracting mainly along the field lines so as not to compress the field to the same extent. The protostar is at first needle-shaped and parallel to the field and its rotational axis is also parallel to the field so as to limit angular momentum; later it contracts to form a disk.

During contraction the stellar magnetic field is severed from the main field by being pinched into an hour-glass form and then pinched so severely that field lines passing through the edge of the disk-shaped protostar form a magnetic neutral sheet in the equatorial plane. Field lines then reconnect across the neutral sheet, so as to convert the hour-glass field to a dipole field (Mestel and Strittmatter, 1967). The various reconnection processes of Section 4.2 may prove important at this stage.

The final problem involving the magnetic field is the removal of excess angular momentum from the star. This problem was discussed in Section 10.1 in connection with the origin of the solar system. It extends to the more general problems of star formation and of why early-type stars rotate faster, on the average, than late-type stars. It is fairly certain that the answer lies in magnetic braking, caused by a magnetic field which links the star with a gaseous disk or cloud. The field lines are deformed and transfer angular momentum outwards from the faster-rotating star. This mechanism is discussed in more detail below and again in Chapters 12 and 13 in connection with galactic central systems and radio galaxies.

(3) The final stages of star formation begin when the opacity of the whole cloud is high enough to stop isothermal collapse. The central temperature increases until the gas pressure and magnetic pressure satisfy the virial equation given in the following section. At this stage Helmholtz contraction of individual protostars begins: gravitational energy is steadily converted to heat and radiated away, thus allowing slow contraction. The condition for this type of contraction in the protocluster (and in the smaller gas clouds) is that the optical depth from the surface to the center attains a value given by

$$\tau > \frac{c}{v_f} \cdot \frac{p_r}{p} \tag{11.3}$$

where p_r and p are the radiation and gas pressures, and c and v_f are the velocities of light and of free fall under the influence of gravity alone. Protostars collapse until their H and He are completely ionized; later a second Helmholtz contraction starts, leading to the generation of nuclear energy and the onset of stellar evolution.

11.3 Magnetic Stars

In 1889 astronomers conceived for the first time the idea that a star (the sun) might possess a magnetic field. At the time there was no way of testing this idea, but only a few years later the Zeeman effect was discovered and in 1908 Hale used this effect to prove the existence of magnetic fields in sunspots. The more difficult task of detecting magnetic fields in stars was not undertaken with success until Babcock (1951) developed equipment of sufficient sensitivity. Since then, some hundreds of stars have been investigated and a large proportion are found to have magnetic fields ranging in strength up to several thousand gauss. These fields vary in strength and, in addition to the question of the origin of the fields, they raise others of an electromagnetic nature concerning the interaction between the thermal-gravitational field and the magnetic and rotation fields.

Origin of the Fields

Following the discussions of the two preceding sections, it is accepted that a magnetic field is frozen into the protostar, so that the origin of the field provides no problem. Nor need the maintenance of the field necessarily provide a problem if we accept Wrubel's (1952) estimate of the e-folding time for the slowest-decaying dipole component of a solar field. This time is 4×10^9 years, which is about the age of the solar system, so that the field could possibly be a fossil field. The discovery of reversal of the solar field seems to rule out this simple explanation in the case of the sun, but it may apply for some stars. Alternatively, magnetic amplification or magnetic dynamo action, discussed in Section 4.1, may be invoked.

Interaction Between Fields

Interaction between the various stellar force fields is described broadly by the virial theorem of Chandrasekhar and Fermi (1953). For a star which is neither expanding nor contracting significantly we have

$$2K + W_m + 3(\gamma - 1) W_h + \Omega = 0 \qquad (11.4)$$

where K is the macroscopic kinetic energy (mainly rotational), W_m is the magnetic energy, W_h is the total heat energy (all integrated throughout the body of the star), γ is the usual ratio of specific heats and Ω is the gravitational energy (negative). If the magnetic field has a substantial external component, then a surface integral should be added to take account of this. Since all of the terms of equation 11.4 except Ω are positive it is evident that the field, like thermal pressure and macroscopic motions, is disruptive. An immediate extreme upper limit of field is found by putting $W_m = - \Omega$ when, for an A star, the limit is 3×10^7 gauss. The virial theorem also provides a very simple relationship between the total magnetic flux through a star ψ and the mass. As shown in Chapter 13, the upper limit of the flux is given by

$$M \sim 10^3 \psi \qquad (11.5)$$

M being in grams and ψ in gauss cm^2.

The scalar virial theorem hides the essential anisotropy of magnetic and centrifugal forces and the limiting field strength and flux are considerably lower than the above values. In a way the magnetic field decreases the stability of a star by increasing its period of radial oscillation; in fact, if equation 11.5 is not satisfied, expansion continues to infinity. However, long before this stage is reached the anisotropic, and perhaps highly localized, magnetic stresses have overcome gravitational control in one region or another and burst forth. The instability may have the form shown in Figure 11.1, which is also similar to that attributed to developing solar coronal magnetic fields (Chapter 5). In any case the field cannot make the star explosively unstable.

Magnetic Variable Stars

Magnetic variable stars were discovered by Babcock (1951) and have stimulated a great deal of theoretical effort. The major problem is to explain the regular variations, even including reversals, of the fields. Theories take three forms: (1) the oblique rotator, (2) the magnetic oscillator and (3) the hydromagnetic cycle. They have been reviewed by Cowling (1965) and discussed by a number of authors in Cameron (1967); all three meet difficulties.

In model (1) patches of different magnetic polarity are revealed as the star rotates about an axis inclined to the line of sight. This model is attractive because of its simplicity and one might be fairly certain that it does account for some observed variations. However, it meets quantitative difficulties, and fails to explain irregular changes which are sometimes superimposed on magnetic cycles. Model (2) depends on changes in the surface magnetic field of the star caused by mechanical oscillations of various types. The principal difficulties lie in providing polarity reversals in integrated light, and in explaining oscillation periods as long as several days. Torsional oscillations may serve, but only if a number of favorable assumptions are made. Model (3) is analogous to models of solar variations, and offers endless variations of form.

The only star whose surface field may be observed in detail is the sun and here we see variations caused by a hydromagnetic cycle and others caused by rotation. It is likely that these effects are important in many stars; some observational evidence is provided by the flare stars mentioned below.

Finally, there is the problem of the apparent absence of a strong surface field in the majority of stars, and of a strong primeval field emanating from the central body of the sun. According to the theory of star formation there should always be a strong frozen-in field, unless some HI clouds happen to be non-magnetic. Other factors, associated with stellar evolution, are discussed by Mestel (in Cameron, 1967).

Magnetic Braking

A vitally important function of stellar magnetic fields is that of magnetic braking. The basic idea, due to Alfvén, is that angular momentum can be transmitted outwards along the lines of force and dispersed into the surrounding interstellar medium. Lüst and Schlüter (1955) proposed a model in which the star had a magnetosphere with a well-defined boundary, such as that sketched in Figure 10.2 as Jupiter's magnetosphere. Frictional forces at the surface brake the rotating magnetosphere and the star and may in this way explain why stars of late spectral type generally rotate slower than early-type stars. In this and a later modification (Cowling, 1965) the stellar field is confined within a well-defined shell and the field lines are bent into spiral forms as shown by the line Q of Figure 10.2. The consequent shearing stress is then

$$p_s = \frac{B_1 B_2}{4\pi} \tag{11.6}$$

where B_1 and B_2 are the field components in, and perpendicular to, the meridian planes. At the surface of the magnetosphere the external pressure p_m balances the magnetic pressure so that

$$p_m = \frac{B_1{}^2 + B_2{}^2}{8\pi} \tag{11.7}$$

The external pressure is due mainly to turbulence and may be estimated. Cowling finds that a star similar to the sun would be brought to rest in a period of about $10^{12} B_0{}^{-5/4}$ years. For a field of 1 gauss (the sun's present field) the effect is negligible, but for a field of 1000 gauss (as in some stars), the braking would take only about 10^8 years.

Schatzman (1962) has suggested that streams of particles being ejected from the star (the stellar wind) would tend to conserve their angular momentum and so would lag behind the star in rotation and exert a braking effect. The loss of only a small part of the mass of the star could remove a large part of the angular momentum and if many stars eject material, then this effect might be dominant. This braking model has been developed in considerable detail by Mestel (1966, 1968), Ferraro and Bhatia (1967) and others.

Other Effects in Magnetic Stars

Other effects which must occur in magnetic stars are impediment of the convective transfer of energy (the basis of theories of sunspots) and surface distortions caused by the presence of magnetic fields. It might seem that one or both effects might greatly change the shape and brightness distribution over a star, but this is not so. Magnetic forces cannot appreciably distort a star unless the internal field is enormously stronger than the surface field. Near the surface magnetic fields may appreciably interfere with surface convection of heat and may produce "starspots" but not large-scale brightness variations. The field may affect convection only in a relatively thin layer and if part of this layer is more "opaque," then the temperature rises below this part to equalize the surface temperature.

These problems are discussed in more detail in the references given above and by Ledoux and Renson (1966) and various authors in Lüst (1965).

11.4 Galactic Radio Sources and x-Ray Sources

In this section we discuss the various sources of radio and x-ray emission lying within our galaxy. These comprise a variety of stellar and non-stellar objects and reveal a number of interesting electrodynamic effects. For convenience the sources are separated into three categories: general galactic radio emission, miscellaneous discrete radio sources and x-ray sources.

General Galactic Radio Emission

General galactic radio emission is mainly synchrotron emission from cosmic-ray electrons gyrating in the galactic magnetic field and has been reviewed by Mills (1964). Synchrotron radiation is received most strongly from the Milky Way, and particularly from the general direction of the galactic central region. However, appreciable radiation is received from all directions and there is some evidence that the galaxy (like some other galaxies) has a corona or halo of magnetic field and cosmic rays. This corona is not flattened like the disk, but is more or less spherical in shape, with a diameter equal to the disk diameter.

From the meter-wave brightness distribution, models of the galactic magnetic field and cosmic-ray electron distribution may be constructed. However, in such models there are two principal variables, the field strength B and electron flux, and one of these must be determined by other means. If the value of B adopted above is used, then the density of cosmic-ray electrons in the general direction of the galactic center is several times its value near the sun. A third variable in any detailed model is the absorption by the HII clouds. These do not contribute much to the meter-wave radiation (because of their relatively low gas temperature) but they may cause significant absorption of synchrotron radiation which passes through them.

Another part of the general galactic radio emission is the line emissions of atoms and molecules, notably that of neutral hydrogen atoms at a wavelength of 21 cm. A review of the theory and observations of this and other lines has been given by Kerr (1968), who gives earlier references to the use of these observations in determining the distribution of hydrogen throughout the galaxy. Observations of line emission have the great advantage that Doppler shifts may be measured and line-of-sight velocities of the gas determined. Since the general pattern of galactic rotation is known, this has allowed estimates to be made of the distances of HI distributions, and a model of galactic HI to be determined.

Not all neutral hydrogen fits into this general rotational pattern, because some HI clouds seen at high galactic latitudes are falling towards the galactic plane with speeds of about 100 km/sec, while others are receding. Oort (1966) has discussed the falling clouds, which may be gas ejected from the central region by an explosion which occurred a few million years ago. This possibility is considered further in Chapter 12.

Miscellaneous, Discrete Radio Sources

Miscellaneous, discrete radio sources comprise HII clouds, OH clouds, supernova remnants, the galactic central region and flare stars.

The HII clouds are strung out along the Milky Way, many being well-known optical objects such as the Omega Nebula. They have been discussed in Section 11.1 and their emission is thermal and well understood. Discrete sources of OH line emission show many peculiar features which are not understood; the interested reader is referred to a review by Robinson and McGee (1967).

The radiation from supernova remnants (excluding the Crab Nebula) shows a shell-like structure and a synchrotron-type spectrum. It appears to be explicable in terms of compression of the interstellar magnetic field and electron cosmic-ray gas outside the expanding supernova shell. Models derived by van der Laan (1962) and Kulsrud et al. (1965) indicate a compression of the external field and gas by a factor of about two. In addition, the cosmic rays suffer betatron acceleration which also increases their synthrotron emission at meter wavelengths. All of these factors combine to provide a discrete radio source. Kulsrud et al. also give an interesting discussion of the Rayleigh-Taylor instability at the boundary, where inertial force exerted by the shell plasma is balanced by magnetic and cosmic-ray pressures. This instability does appear to be very widespread.

Observations of the central region have been discussed by Burke (1965) and by Kerr (1967). It has a complex pattern of thermal, non-thermal and line emission, the latter comprising the 21 cm line of neutral hydrogen, the 18 cm group of OH lines and a hydrogen recombination line at 6 cm. It is not easy to separate synchrotron and thermal emission but the 21 cm line measurements do give a physical picture of the neutral hydrogen density and velocity distribution and this is sketched in Figure 11.2 with the plane of the paper in the galactic plane. In the center (within the 1 kpc ring) is a central core having mainly rotational motion, and a ring with some radial motion. Next we have a bar which joins the so-called 3 kpc arm, the latter having mainly rotational but some radial motion. The gas layer is thin, about 100 pc between half-density points. It has been suggested that motions

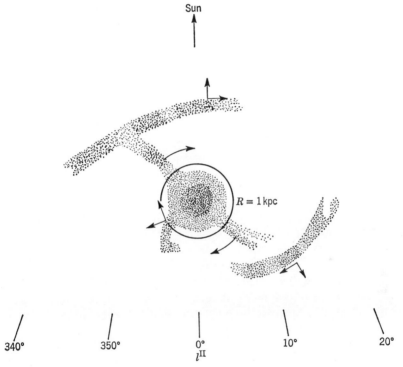

Fig. 11.2. A possible structure of the galactic-center region (after F. J. Kerr, 1967).

in the central region have resulted from a central explosion, but this would not explain the high angular velocities observed, because gas ejected explosively would tend to preserve angular momentum. As seen in Chapter 12, it is likely that electromagnetic forces play a major part.

Although discrete radio sources were first called radio "stars," it was found that the optical stars in general provide little or no detectable radiation. The sun, of course, is seen to emit thermal radiation and a variety of non-thermal emissions. However, these are only detectable because of the sun's proximity. A few "flare stars" emit enough radiation to be measured; the mechanisms responsible may be similar to those on the sun, but much more powerful. The main difficulty involved in obtaining flare-star data is the time required to keep a continuous watch on a few stars, which may not produce flares in any case. This work on UV Ceti has been described by Higgins *et al.* (1967), who give some earlier references.

x-Ray Sources

Galactic x-ray sources comprise a variety of objects, some of which are certain to involve interesting electrodynamic processes. The first discovered (in 1962) is Sco X-1, and for photon energies above 10 keV its power output of about 5×10^{36} erg/sec exceeds that of the quiet sun by more than 10^{14}. Clearly the sun, like most other stars, is not a notable x-ray source and some galactic sources have x-ray emission 10^4 times the total power output of the sun at all frequencies. Some dozens of sources have been located, most concentrated near the galactic plane and therefore lying within the galaxy. Their properties have been reviewed by Morrison (1967), Friedman (1968) and Oda (1968).

One source positively identified with an optical (and radio) object is the Crab Nebula, discussed in the following section. Another is the strongest source Sco X-I, which is a blue star, fluctuating in brightness as much as one magnitude in a day and distant from the earth a few hundred parsecs. This source emits 1000 times more x-ray energy than optical energy. A few other sources have been identified with optical objects with varying degrees of certainty. These objects are supernova shells, a peculiar gaseous nebula, another blue star and an external galaxy M87. The general distribution of x-ray sources seems to coincide with that of young stars, and so of spiral arms which these stars define. Radiation from normal stars other than the sun, and from normal galaxies, has not been detected.

There appear to be at least four types of x-ray sources, one of which (the Crab Nebula) is discussed in the following section and another (a peculiar galaxy) in Chapter 13. A third type is that associated with supernova shells, particularly Cas X-I, which is also a strong radio source. All radiation may be explained by synchrotron emission of compressed galactic field and cosmic-ray electrons. Gould's (1965) Compton-synchrotron process may enhance x-ray emission by collisions between lower-energy (synchrotron) photons and the cosmic-ray electrons. This source type might account for the association with young stars, which provide the most violent supernovas.

A fourth type of x-ray source is required to explain the rapid variations in intensity which have been observed (Lewin *et al.*, 1968) and which indicate dimensions of 1 AU or less. A likely mechanism is that of Rossi, of bremsstrahlung from a cloud of gas of mass about 10^{-5} $M_{\odot}$ at a temperature of about 5×10^{7} K. The difficulty is to prevent such a gas cloud from exploding; even for the maximum permissible extent a magnetic field of some hundreds of gauss would be required. Shklovskii has suggested gravitational control: gas from a close binary companion is accreted by a neutron star, gaining gravitational energy of about 10^{20} erg/g or about 10^{7} eV per particle. An alternative model, also based on a neutron star, has been suggested by Cameron (1965). This uses the vibrational energy of the star and its powerful magnetic field to radiate hydromagnetic waves outwards. Gas at the surface and in the magnetosphere is heated and radiates by the bremsstrahlung and synchrotron mechanisms.

11.5 The Crab Nebula

The Crab Nebula (optical object NGC 1952, radio object Tau A, x-ray source Tau X-1) is a galactic nebulosity unique in appearance and in a number of other respects. The optical object so impressed Messier that he listed it as M1; it is also of historic interest, originating as a supernova observed by Chinese astronomers in 1054 A.D. It is the only example, apart from some scattered fragments, of a galactic Type I supernova. This type involves the explosion of an old and chemically complex star from which a gaseous shell of mass less than 1 $M_{\odot}$ is ejected with velocity about 10^{3} km/sec. Electromagnetic radiation from the Crab, having a variety of mechanisms of emission, has been observed from radio to gamma-ray frequencies. The nebula is also a source of cosmic rays and plays a major part in the "supernova theory" of primary cosmic radiation (Section 12.5). It probably also contains a mechanism for the creation of magnetic field—a magnetic amplifier (Section 4.1).

In this section we first discuss the observational data relating to the Crab Nebula, and then two rival models which attempt to explain the observations.

Observations

Observations of the Crab Nebula were at first confined to the optical spectrum, but even these proved inexplicable until Shklovskii

(1953) suggested their origin was by the synchrotron process. This was soon demonstrated by showing plane polarization of the light. The optical emission has revealed two peculiarities of the nebula which we consider most significant in connection with the origin of all types of radiation and of electrodynamic processes in general. The first is the observation of luminous whisps moving outwards from the center of the nebula with speeds of about 3×10^9 cm/sec. Since the emission is synchrotron, these must be caused by localized enhancements of cosmic-ray electron flux or of magnetic field strength or both. Another most interesting discovery, made by Baade (1942) and confirmed by Trimble (1968), is that the plasma shell is not decelerating outwards as might be expected, but is actually accelerating at an average rate of 10^{-3} cm/sec^2. These and other effects indicate a continuous generation of cosmic rays and probably also magnetic energy in the central region.

The shell of the Crab Nebula is either a prolate or oblate spheroid, again possibly indicating an internal accelerating force which is anisotropic in form. Detailed arguments (Shklovskii, 1966; Trimble, 1968) suggest a prolate form and a distance of 1.5–2.0 kpc. The dimensions (angular extent 4' by 3') are then 1.7–2.3 pc by 1.3–1.7 pc and the volume is $5.3–9.4 \times 10^{55}$ cm^3.

The Crab is bright at all frequencies from about 10^7 c/sec to above 10^{20} c/sec (30 m to 0.02 Å). The luminosity is 2×10^{34} erg/sec in the radio range and increases to 2×10^{36} erg/sec in the optical range and 2×10^{37} erg/sec in x rays. This radiation is mainly synchrotron, and more recently Haymes et al. (1968) have shown that gamma radiation with energy quanta up to 560 keV (0.022 Å) fits smoothly onto the x-ray spectrum. The total power radiated up to the gamma radiation limit of observations is a few times 10^{37} erg/sec, so that the Crab may be regarded as primarily an x-ray and gamma-ray source. In addition to all the synchrotron radiation, radio emission at 7.9 m is provided by a small central source, only about 2×10^{-3} pc in diameter (Bell and Hewish, 1967). This source provides about 20% of the emission at 7.9 m and is too bright to be accounted for by incoherent synchrotron emission (Shklovskii, 1966). It may be accounted for by coherent synchrotron emission or by space-charge waves (Section 4.5), either of which requires the presence of plasma of density about 10^5/cm^3 or more.

The brightness distribution over the Crab Nebula differs widely for different spectral regions; it is plotted in one dimension in Figure 11.3 for four spectral regions. Curve a corresponds to a simple model (a circle of diameter 1'.7) which fits the x-ray observations. Curve b

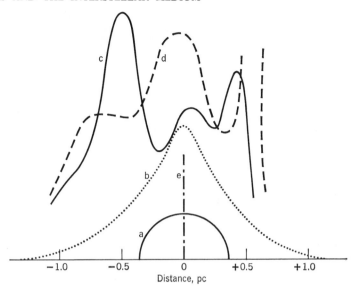

Fig. 11.3. The brightness distributions in one dimension across the Crab Nebula—a somewhat schematic representation. Curve a: The idealized x-ray brightness distribution. Curve b: The optical distribution. Curve c: The 1.4 m radio distribution. Curve d: The 5.0 m radio distribution. Line e: The 7.9 m narrow-beam radio disturbance.

shows the optical brightness distribution, while curves c and d correspond to radio observations made at 1.4 m and 5.0 m (Artyukh *et al.*, 1966). The vertical line e denotes the tiny central source, isolated by high-resolution techniques. Measurements by Gardner (1965) at 6 cm show a central maximum and gradual decrease outwards. Summing up these data we conclude that there are two "central" sources and one ring or shell source. The synchrotron emission at 6 cm, optical and x-ray wavelengths show increasing concentration towards the center with decreasing wavelength. The non-synchrotron source (line e) is a tiny, discrete central object. Other meter radio emission (synchrotron) is concentrated towards the shell of the nebula.

These observational data provide three main problems. First, the synchrotron losses require an input, at the present time, of cosmic-ray electrons amounting to more than 10^{38} erg/sec. These must originate in the central region where the only obvious energy source, radioactive decay of the supernova remnant, fails by a factor of about 100 (Clayton and Craddock, 1965). The total energy requirement in the form of fast electrons and perhaps fast ions may exceed 10^{49} erg, which is more than the kinetic energy provided by the supernova explosion.

Second, the synchrotron emission requires a magnetic field of strength $B \sim 2 \times 10^{-4}$ gauss and total energy about 10^{47} erg. The polarization measurements indicate a large-scale, ordered field to which we may attribute a magnetic flux $\psi = BL^2$, where L is a characteristic dimension. For $L = 1$ pc, the flux is 2×10^{33} gauss cm^2 and as seen in Section 11.3 this could not have passed through the central star prior to its explosion. Two alternative sources of the magnetic flux have been suggested, one outside the expanding shell and one by a magnetic amplifier in the central region; these are discussed below.

Third, we have the problem of the central radio source, which probably radiates by the interaction of plasma oscillations with the magnetic field. The source is notable at a wavelength of 5 m so that the electron density must be at least 4×10^7/cm^3. The source size is 10^{-3} pc and if the electron and proton densities are uniform throughout, the mass of the HII cloud is about 10^{-3} M$_\odot$.

Galactic Magnetic Field Model

A galactic magnetic field model (Kulsrud *et al.*, 1965) is based on compression of the galactic field by the expanding supernova shell, and subsequent penetration of some of the compressed field through the shell as a result of the Rayleigh-Taylor instability. This model does not appear to meet the energy requirement of about 10^{49} erg. Observational and theoretical estimates of the shell mass agree on a value less than 1 M$_\odot$ and with its present velocity of 10^8 cm/sec the total shell kinetic energy is less than 10^{49} erg. Since the shell has if anything accelerated, its kinetic energy could not be a major source of energy; even if it were, most would be transferred outside the shell and not into the central region.

This model seems to explain the ring-type radio brightness distribution of some other supernova remnants. In so doing it fails for the Crab Nebula where the main activity is near the center of the nebula. It is probable that all three of the above problems must be solved together in terms of some mechanism in the central region. This cannot be operated by the products of radioactive decay, and as well as cosmic-ray particles it must account for the magnetic flux of the nebula. It is likely that the observed outward moving, luminous whisps are due to outward moving magnetic field lines and trapped cosmic-ray electrons. The observed concentration of meter-wave radio emission in the outer regions may then be explained by the adiabatic losses of the electrons and the decrease in the frequency of peak synchrotron emission from gamma rays (10^{14} eV) to meter radio waves (10^9 eV).

Spiral Magnetic Field Model

A spiral magnetic field model of the Crab Nebula (Piddington, 1957, p. 530) was based on a spinning central object connected to the shell by magnetic field lines, and so acting as a magnetic amplifier (Section 4.1). The energy source proposed was that of radioactive decay, but this now appears inadequate. However, the discovery of a radio and optical pulsar in the center of the Crab Nebula (Section 11.6) and the development of pulsar theory, makes it likely that the central star is a neutron star rotating at a rate of 30 rev/sec. If this star has a mass of about one solar mass and radius about 10 km, then its kinetic energy of rotation is about 10^{49} erg. This is a

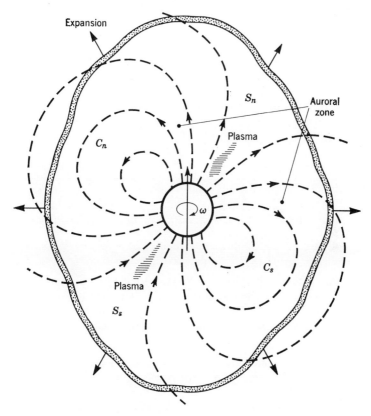

Fig. 11.4. A highly schematic model of a young supernova, with an outward expanding shell (dotted) and a collapsing star. All magnetic field lines emerging from the star originally linked with the shell, but some have broken away to form a corotating magnetosphere (C_n, C_s). Some remain linked and are separated by "auroral" zones.

thousand year supply and so is adequate, as is the angular momentum.

In Figure 11.4 we see the model in an early stage of development, very shortly after the supernova explosion. The collapsing star and expanding shell are linked by magnetic field lines shown dashed. All field lines emerging from the original star are initially linked to the plasma shell, but as the star shrinks most of these are likely to reconnect in the manner discussed in connection with star formation (Section 11.2). In Figure 11.4 the dipole type stellar field is oblique to the rotational axis ω and so the reconnected lines in the meridian section shown form a northern corotating magnetosphere (C_n) and a southern corotating magnetosphere (C_s). The whole corotating magnetosphere is shaped as a tilted doughnut. Some polar field lines must remain connected to the shell and so we have northern and southern shell-connected fields, S_n and S_s, separated from C_n and C_s by "auroral" zones.

Now as the star shrinks it tends to conserve angular momentum and so spins faster. The field lines S_n and S_s are wound into helical and spiral forms and magnetic amplification is caused by shear motions $V(r, \theta)$ in the plasma. In an r, θ, ϕ system of axes the amplification is given by the frozen-in equation (2.13), which expands to

$$\frac{\partial B_\varphi}{\partial t} = \sin \theta \left(B_\theta \frac{\partial \omega}{\partial \theta} + r B_r \frac{\partial \omega}{\partial r} \right) \tag{11.8}$$

If a fraction δ of the total flux ψ_s through the star remains connected to the shell, or to plasma known to pervade the whole nebula, then at each revolution of the star the nebular flux ψ increases by an amount $\delta\psi_s$, so that after n revolutions

$$\psi = n\delta\psi_s \tag{11.9}$$

The present rate of rotation is 30/sec and if the average is taken at, say, twice this value, then $n \sim 2 \times 10^{12}$. The present nebular flux is 2×10^{33} gauss cm^2, so that $\delta\psi_s \sim 10^{21}$ gauss cm^2 (about that of an average sunspot). A main sequence star with a reasonable surface field of 2 gauss has $\psi_s \sim 10^{23}$ gauss cm^2, so that $\delta \sim 10^{-2}$ and only 1% of the original flux need remain connected to external plasma. If this is the polar cap field, then the "auroral" zones occur at colatitudes $\theta \sim 7°$ and $173°$.

The model would seem to explain, in a very simple manner, a number of features of the Crab Nebula. The large magnetic flux results simply from the winding-up process and the resulting outward magnetic pressure may explain the outward moving light whisps and the shell acceleration and shape (a prolate ellipsoid forms as a toroidal field expands with its plasma). If magnetic flux is created at a faster rate than that found above, and some is destroyed by magnetic annihilation in the spiral neutral sheet, then the magnetic energy destroyed will be converted to particle energy and may account for some of the cosmic-ray electrons.

Finally, we should note that, while the direct source of magnetic energy is kinetic, the original source is gravitational energy of stellar collapse. Here is strong evidence for the conversion of gravitational energy to magnetic energy, as required for models of spiral and radio galaxies and quasars discussed in Chapters 12 and 13.

11.6 Pulsars

Pulsars are galactic radio sources which pulsate rapidly and with remarkable regularity, with repetition periods between about 0.03 sec and a few seconds. They are best observed at frequencies between a few tens and a few hundreds of megahertz. The first pulsar observed, CP1919 (Hewish *et al.*, 1968) emits pulses lasting a fraction of a second with a period about 1.337,301 sec. At first it was thought that the time-keeping accuracy might equal or exceed that of our best clocks, but later measurements of Davies *et al.* (1969) showed a slowing down. The increase in pulse period p per period is given by $dp/p = (1.1 \pm 0.5)$ 10^{-15}, and several other pulsars show comparable increases of a few parts in 10^8 per year. The pulsar in the Crab Nebula (Comella *et al.*, 1969) is slowing down at the rate of approximately 1 part in 2400 per year, and it is interesting to note that this pulsar has the shortest period (33 msec) and is probably the youngest, being only about 900 years old. It may be that age, period and rate of slowing of pulsars are generally related in this way.

The impulsive nature of the signal recorded from CP1919 results from the periodic passage of a signal of limited frequency range and descending frequency through the bandpass of the receiver. As we have seen, this period is extremely regular. On the other hand, all other characteristics of the signals are most irregular: the intensity fluctuates, the pulse shape varies and the polarization of a given source may be linear or circular. These characteristics, together with the pattern of slowing down, must be accounted for in any adequate theory of pulsars.

Estimates of pulsar distances may be made from the observed dispersion of the pulse components. If the emitted pulse is a perfect transient with no dispersion, then the observed frequency drift, caused by the lower velocities of the lower-frequency components in the interstellar gas, is given by

$$\frac{\partial \omega}{\partial t} = - \frac{c}{L} \frac{\omega^3}{\omega_p{}^2} \qquad (11.10)$$

where ω_p is the plasma frequency and L is the path length. If the interstellar electron density is taken as $0.01/\text{cm}^3$, then distances so determined range from a few hundred to a few thousand parsecs. One pulsar has been seen in HI absorption, indicating that it lies behind a hydrogen cloud thought to be at a distance of about 4 kpc. Another pulsar may be identified with the extended radio source Vela X (Large *et al.*, 1968) which, in turn, is identified with a supernova remnant at a distance of about 500 pc.

The durations of the pulses and of their substructure allow estimates to be made of the upper limits of the sizes of the emitting regions. The pulses exhibit great complexity, with characteristic times, t, of the main pulse and substructure ranging from a few times 10^{-2} sec to a few times 10^{-6} sec. The corresponding limits to the sizes, ct, of the sources responsible are about 10^4–1 km, which suggests stellar, or smaller, objects.

In spite of the small source size, the power radiated by pulsars is comparable with that of other notable galactic sources. On the assumption of a distance of 4 kpc and isotropic radiation, the source CP0328 radiates about 10^{31} erg per pulse, with a peak power of about 3×10^{33} erg/sec. This is about 10% of the radio power of the Crab Nebula and 10^{11} times larger than that of a great solar radio burst. The radio brightness temperature in a pulsar must be enormous (far above 10^{20}K), which requires a coherent process of organized plasma oscillations or organized synchrotron emission. In addition to this remarkable radio emission, optical astronomers at Steward Observatory (Cocke *et al.*, 1969) and later at McDonald Observatory and Kitt Peak National Observatory have detected an optical pulsar in the Crab Nebula. Its flashes keep precisely in step with the radio pulses.

The most significant features of pulsars, to be explained by any adequate theory, is the extreme regularity and fast repetition of the pulses, together with the irregularity of all other properties. The timekeeping requirement seems to imply objects which rotate, orbit or vibrate as a whole. The fast rate requires small objects and only neutron stars seem capable of satisfying the requirement. These are hypothetical products of supernova explosions, with degenerate nucleon gas interiors which may vibrate with periods about 1 msec to a maximum in the vicinity of 1 sec. Thorne and Ipser (1968) conclude that neutron stars with this long period would probably be energetically unbound, but that further theoretical study is desirable. Rotational and orbital models involving neutron stars may have periods as short as 10^{-3} sec and so easily satisfy the period requirement.

The various models and the difficulties they meet may now be listed.

(1) The orbiting models face the possibly fatal objections of gravitational radiation with consequent change in period, and of gravitational tidal effects. They may also meet some of the beaming difficulties of the rotating models, mentioned below.

(2) A pulsating star is an excellent timing device and avoids the stringent beaming requirements of the orbiting and rotating models. However, current theory of neutron stars seems unable to account for the longer pulse periods, or for the observed increases of pulsar period.

(3) Rotation of a star provides an excellent clockwork mechanism, and if the star is a neutron star (radius about 10 km), then it may possess a powerful magnetic field, developed simply by the compression of an ordinary stellar field and conceivably as high as 10^{10}–10^{14} gauss (Woltger, 1964; Cameron, 1965). The existence of a stellar magnetosphere obviously raises many possibilities of providing radio emission, and the observed decrease in pulse rate may be accounted for simply by magnetic braking of the star. In fact the observed decrease in the rate for the Crab pulsar would provide the right amount of energy to account for all of its emissions.

The difficulties met by rotational models are the requirement for a narrow beam, the emission spectrum and some other observational data discussed below. A "localized emitting region" on a star (Ostriker, 1968) does not necessarily provide a narrow beam, as we have seen in the case of solar flares whose radio emission is spread over most of a hemisphere. An interesting model based on a rotating neutron star with an oblique dipole magnetic field (Pacini, 1967) was actually proposed before the discovery of pulsars had been announced. If q is the projection of the dipolar moment on the plane perpendicular to the axis of rotation, and ω is the angular velocity, then there will be a monochromatic emission of electromagnetic waves of frequency ω and intensity

$$I = \frac{2}{3} q^2 \omega^4 c^{-3} \qquad (11.11)$$

For the apparent rate of rotation of the Crab and a surface field of 10^9 gauss, $I \sim 10^{32}$ erg/sec which is about the order required by pulsars. However, in other respects this emission bears no resemblance to that of a pulsar and its interest lies in the interaction of the radiation with the surrounding plasma and magnetic field and with the braking effect on the star (Gunn and Ostriker, 1969). The latter may impose an upper limit on the moment q, since too much braking would result in an excessive decrease of pulse rate.

(4) A more promising rotational model (Gold, 1969) depends on the fact that plasma in a corotating magnetosphere may attain velocities approaching that of light. This occurs at a distance $r_c \sim c/\omega$ from the rotational axis, or about 1600 km in the case of the Crab. The hypotheses are made that plasma is ejected from the star from one or two highly localized regions and that in the region of relativistic velocities electrons and positive ions separate to form electric space charge clouds. These provide coherent radio emission by virtue of their acceleration and are then flung out of the corotating region.

The model is conceptually simple and attractive, but meets several difficulties. The large ratio of pulsar period to pulse width imposes severe beaming conditions and requires very highly localized and rigidly positioned "star spots" from which plasma emerges. The narrow "lighthouse" beam must remain little changed over a wide frequency spectrum, yet a simple pencil beam seems unlikely because few pulsars would be detectable from the earth. In the slower pulsars the critical radius r_c must exceed 10^5 km or 10^4 stellar radii and the magnetic field will be reduced by a factor of 10^{12} below its surface value and may be too weak to give corotation.

(5) So attractive is the rotating neutron star as a clock mechanism that an alternative model, not dependent on star spots or relativistic effects, is worth seeking. Such an alternative may, perhaps, be a part of the Crab magnetic field model discussed in Section 11.4 and shown in an early stage of development in Figure 11.4. Here we have a co-rotating magnetosphere which is asymmetrical about the rotational axis. There is also a plasma and field configuration which does not rotate with the star and which is also asymmetrical. Between the rotating and non-rotating regions the field and plasma must be constantly accelerated back and forth, so as to satisfy the hydromagnetic force equation 2.6 everywhere throughout the rotating and non-rotating regions. At one or more positions of the rotating magnetosphere the velocity of some plasma exceeds the Alfvén velocity and a shock wave is set up. As in the case of solar type II bursts, this shock wave will give rise to more or less omnidirectional radio emission and this is the pulsar emission. The synchronous optical emission from the Crab is simply enhanced synchrotron radiation caused by the compression of the magnetic field and cosmic-ray gas.

This is a true pulsar, or pulsating source, although the pulsations are driven by the rotating star and magnetosphere (Piddington, 1969).

References

Artyukh, V. S., Vitkevitch, V. V., Vlasov, V. I., Kafarov, G. A. and Matveenko, L. I., 1966, *Soviet Astron. AJ* **10**, 9; *Astron. Zh.* **43**, 13.

Baade, W., 1942, *Astrophys. J.* **96**, 188.

Babcock, H. W., 1951, *Astrophys. J.* **114**, 1.

Bell, S. J. and Hewish, A., 1967, *Nature (London)* **213**, 1214.

Burke, B. F., 1965, *Ann. Rev. Astron. Astrophys.* **3**, 275.

Cameron, A. G. W., 1965, *Nature (London)* **205**, 787.

Cameron, R. C., ed., 1967, *The Magnetic and Related Stars,* Mono Book Corp., Baltimore.

Chandrasekhar, S. and Fermi, E., 1953, *Astrophys. J.* **188**, 116.

Clayton, D. D. and Craddock, W. L., 1965, *Astrophys. J.* **142**, 189.

Cocke, W. J., Disney, M. J. and Taylor, D. J., 1969, *Nature (London)* **221**, 525.

Comella, J. M., Craft, H. D., Lovelace, R. V. E., Sutton, J. M. and Tyler, G. L., 1969, *Nature (London)* **221**, 453.

Cowling, T. G., 1965, *Stars and Stellar Systems* **8**, 425, University of Chicago Press, Chicago.

Davies, J. G., Hunt, G. C., and Smith, F. G., 1969, *Nature (London)* **221**, 27.

Davies, R. D., Booth, R. S., and Wilson, A. J., 1968, *Nature (London)* **220**, 1207.

Ferraro, V. C. A. and Bhatia, V. B., 1967, *Astrophys. J.* **147**, 220.

Friedman, H., 1968, *Nature (London)* **220**, 862.

Gardner, F. F., 1965, *Aust. J. Phys.* **18**, 385.

Gardner, F. F. and Whiteoak, J. B., 1966, *Ann. Rev. Astron. Astrophys.* **4**, 245.

Gold, T., 1969, *Nature (London)* **221**, 25.

Gould, R. J., 1965, *Phys. Rev. Letters* **15**, 577.

Gunn, J. E. and Ostriker, J. P., 1969, *Nature (London)* **221**, 454.

Haymes, R. C., Ellis, D. V., Fishman, G. L., Kurfess, J. D. and Tucker, W. H., 1968, *Astrophys. J.* **151**, L9.

Hewish, A., Bell, S. J., Pilkington, J. D. H., Scott, P. F. and Collins, R. A., 1968, *Nature (London)* **217**, 709.

Higgins, C. S., Solomon, L. H. and Bateson, F. M., 1967, *Aust. J. Phys.* **21**, 725.

Kahn, F. D. and Dyson, J. E., 1965, *Ann. Rev. Astron. Astrophys.* **3**, 47.

Kerr, F. J., 1967, *Proc. I.A.U. Symposium No. 31* (*Noordwijk, 1966*).

Kerr, F. J., 1968, *Stars and Stellar Systems* Vol. 7, University of Chicago Press, Chicago.

Kulsrud, R. M., Bernstein, I. B., Kruskal, M., Fanucci, J. and Ness, N., 1965, *Astrophys. J.* **142**, 491.

Large, M. I., Vaughan, A. E. and Mills, B. Y., 1968, *Nature (London)* **220**, 340.

Ledoux, P. and Renson, P., 1966, *Ann. Rev. Astron. Astrophys.* **4**, 293.

Lewin, W. H. G., Clark, G. W. and Smith, W. B., 1968, *Astrophys. J.* **152**, L55.

Lüst, R., ed., 1965, *Stellar and Solar Magnetic Fields*, North-Holland, Amsterdam.

Lüst, R. and Schlüter, A., 1955, *Z. Astrophys.* **38**, 190.

Mathewson, D. S. and Nichols, C. D., 1968, *Astrophys. J.* **154**, L11.

Mestel, L., 1966, *Mon. Not. Roy. Astron. Soc.* **133**, 256.

Mestel, L., 1968, *Mon. Not. Roy. Astron. Soc.* **138**, 359; **140**, 177.

Mestel, L. and Strittmatter, P. A., 1967, *Mon. Not. Roy. Astron. Soc.* **137**, 95.

Mills, B. Y., 1964, *Ann. Rev. Astron. Astrophys.* **2**, 185.

Morrison, P., 1967, *Ann. Rev. Astron. Astrophys.* **5**, 325.

Oda, M., 1968, *Space Sci. Rev.* **8**, 507.

Oort, J. H., 1966, *Bull. Astron. Inst. Netherlands* **18**, 421.

Ostriker, J., 1968, *Nature (London)* **217**, 1227.

Pacini, F., 1967, *Nature (London)* **216**, 567.

Parker, E. N., 1966, *Astrophys. J.* **145**, 811.

Parker, E. N., 1967, *Astrophys. J.* **149**, 517 and 535.

Parker, E. N., 1968, *Astrophys. J.* **154**, 875.

Piddington, J. H., 1957, *Aust. J. Phys.* **10**, 515 and 530.

Piddington, J. H., 1969, *Nature (London)* **222**, 965.

Pikel'ner, S. B., 1968, *Ann. Rev. Astron. Astrophys.* **6**, 165.

Robinson, B. J. and McGee, R. X., 1967, *Ann. Rev. Astron. Astrophys.* **5**, 183.

Schatzman, E., 1962, *Ann. Astrophys.* **25**, 18.

Shklovskii, I. S., 1953, *Dokl. Akad. Nauk.* **90**, 983.

Shklovskii, I. S., 1966, *Soviet Astron. AJ* **10**, 6; *Astron. Zh.* **43**, 10.

Spitzer, L., 1963, *Origin of the Solar System*, R. Jastrow and A. G. W. Cameron, eds., Academic Press, New York.

Spitzer, L., 1969, *Diffuse Matter in Space*, Interscience Publishers, New York.

Thorne, K. S. and Ipser, J. R., 1968, *Astrophys. J.* **152**, L71.

Trimble, V., 1968, *Astronom. J.* **73**, 535.

van de Hulst, H. C., 1967, *Ann. Rev. Astron. Astrophys.* **5**, 167.

van der Laan, H., 1962, *Mont Not. Roy. Astron. Soc.* **124**, 125.

Verschuur, G. L., 1968, *Phys. Rev. Letters* **21**, 775.

Woltjer, L., 1964, *Astrophys. J.* **140**, 1309.

Wrubel, M. H., 1952, *Astrophys. J.* **116**, 291.

CHAPTER 12

Galactic Forms and Activity

In the hierarchy of cosmic objects we move outwards from stars, star clusters and gas clouds to other galaxies such as our neighbor, the great nebula in Andromeda, M.31. Like our own galaxy, these are massive objects made up of stars, gas, dust and the ubiquitous magnetic field. In our local group of about twenty galaxies we have, very fortunately, a fairly full range of sizes and shapes. The largest, and also among the largest in the universe, is M31, with a diameter of about 50 kpc, compared with about 30 kpc for our galaxy. Others range in size down to the insignificant Draco system of diameter 1.6 kpc. Galaxies range in mass up to a fairly well-defined limit of about 10^{12} $M_\odot$; it seems that larger systems cannot form or else they break up.

Galaxies were first observed, visually, as small fuzzy objects, but when they were photographed at the turn of the century it was found that they had a variety of forms and that some had most beautiful and complex structure. This includes bright central systems, spiral arms and occasionally a bridge linking two galaxies. The prototype of this form is M51 which is shown in Figure 12.1; it is seen that one of the pair of spiral arms links M51 to its neighbor. Spirals and other galactic forms are discussed in Section 12.1 and it is probably not too much to say that the understanding of why there are these different kinds of galaxies, including radio galaxies, constitutes the biggest problem in present-day astronomy.

There are many facets to this problem and it is certain that some at least will be understood only when electrodynamic effects are taken into account. A movement in this direction was made with the discovery of a magnetic field in our galaxy and in radio galaxies, and with the spread of the idea of an intergalactic magnetic field. Some attempts were made to explain spiral forms and other effects in this way, but these failed because of the oversimplification of the assumed geometry. Theoretical discussion in this chapter is largely devoted to an extension of these early ideas in the form of a unifying hypothesis. Like other unifying

Fig. 12.1. The beautiful spiral galaxy M51, one of whose arms links it to a smaller companion galaxy (from A. R. Sandage, 1961).

hypotheses, this invokes a single physical quantity to account for many different phenomena. The quantity is the intergalactic field vector and the different phenomena result from the variation in the angle between this (large-scale) field and the galactic rotational vectors. Because it embraces such a wide range of phenomena the hypothesis is highly vulnerable to disproof on both theoretical and observational grounds; such was the fate of the earlier models and, conversely, such is the strength of any surviving model. The various phenomena concerned are listed below.

The way in which *galaxies originate* is not understood, but it is generally accepted that they, like stars and some smaller bodies, form by the contraction of much larger gas clouds. It was perhaps Newton who first conjectured that a homogeneous, unbounded medium composed of gravitating particles would be unstable against gravitation, whereas a finite mass would not. Jeans attempted a mathematical formulation of these ideas in terms of sound waves with a gravitational acceleration added. Von Weizsaecker added the effects of supersonic turbulence to encourage gravitational instability, and Hoyle added the effects of radiative cooling. These and subsequent additions make up a reasonably consistent theory of the origin of a gas cloud of galactic proportions. It is a different matter, however, when it comes to explaining the remarkable variety of *galactic forms*. It does not seem possible to explain these in terms of a simple self-gravitating, rotating gas cloud, or by evolution from one form to another. In Section 12.1 we outline a hydromagnetic theory of the different galactic forms.

Most striking of these forms are the *spirals*. These may have a more or less circular, amorphous central region such as that seen in Figure 12.1 Alternatively the spiral arms may connect to a bar-shaped object which may constitute most of the galaxy; these are called barred spirals. The obvious problem for investigation in these galactic forms is the origin of the spiral arms, and since our galaxy is probably a spiral the theory should be consistent with local optical and radio data. However, an important point to settle before comparing rival theories of the origin of spiral arms, is the nature of the spiral form. Some claim to see a single pair of arms starting from the edge of the structureless central disk and extending throughout the galaxy; others see only spiral bits and pieces or a large number (as great as fourteen) of arms. I believe that the twin arm is the basic structure, but is often obscured by arm branching and by dark, obscuring clouds as in Figure 12.1. This galaxy is neither the best or worst example of an open spiral in which two arms may be traced throughout the disk. In Section 12.2 the twin arm spiral is explained in terms of an electromagnetic force acting on the surface of the disk. Previous hydromagnetic theories failed because they invoked only accelerations parallel with the disk, and these are much too small. However, the ratio of disk diameter to average thickness is about 100 and so motions perpendicular to the disk are that much more effective in promoting gas and star concentrations.

Attempts to locate spiral arms in our galaxy have been made by observing bright young stars and by 21 cm line radio measurements of HI, and are reviewed by Sharpless (1965). Within a few kiloparsecs of the sun both investigations reveal a few concentrations of stars and gas but there is little if any correspondence between the two, and no real evidence of spiral arms. As far as the stellar concentrations are concerned, there are several places in M51 (Figure 12.1) where an observer would see three more or less parallel "arms" which are really only branches of one much more extensive arm. As far as the HI concentrations are concerned, these may have little significance in connection with optical spiral arms, although one often reads of "the two methods of plotting the spiral structure." As seen in Section 12.2, the optical spiral arms comprise young stars and clouds of ionized hydrogen and are probably standing waves of accelerated star formation. While faster star formation may be expected in regions of higher average HI concentration, there are other factors (see Section 11.2) which may dominate. Thus the "two methods of measuring spiral structure" may, in fact, measure different quantities.

The *central systems* of spiral and elliptical galaxies pose a number of problems. First, as Baade pointed out, the density of matter in these regions is far too great to be accounted for in terms of a contracting cloud of roughly uniform density; also the angular momentum per unit mass of this matter is far too small. Second, central systems often have peculiar structural forms including rings and short arms such as those illustrated in Figure 11.2. Third, explosions or outbursts occur in some central regions with the ejection of plasma along the rotational axes. In this respect, spiral galaxies show some of the characteristics of radio galaxies and quasars, which are discussed in Chapter 13. These peculiarities of galactic central regions are discussed in Section 12.3.

The *nucleus* of a galaxy is a small separate central feature which appears in photographs as a stellar or nearly stellar object; its diameter may be 1% or less of the whole central system. The nuclei of some galaxies, called Seyferts, are remarkably active, their spectral lines indicating gas velocities up to 6–8×10^3 km/sec. Even more remarkable is the enormous power radiated in the infrared continuum, amounting perhaps to 10^{60} erg or more during their lifetimes. These galaxies are also closely related to the powerful radio sources, some of which also have remarkable nuclei. The Seyfert galaxies and related objects are discussed in Section 12.4.

The origin of the *cosmic radiation* reaching the earth is not known, but radio astronomy has revealed several possible sources. One possibility is the radio galaxies and quasars and if these prove to be the main source, then this subject should have been considered in the following chapter. However, at present it seems much more likely that the primary cosmic radiation originates within our galaxy, and that spiral systems in general provide their own cosmic rays. The most likely source seems to be the central system, which is often a source of strong synchrotron emission. The relative contributions of cosmic-ray sources are discussed in Section 12.5.

Before entering into these various discussions it should be made clear that by far the greater part of theoretical work in these areas has been carried out with little or no consideration being given to electromagnetic force fields. This work does not directly concern us here except to note that it has had little success in solving the seven major problems listed above. On the other hand, the galactic models determined largely by electromagnetic force fields are speculative and are based squarely on the existence of an intergalactic field. Nevertheless, if such a field does exist, it must be deformed into combinations of helical and spiral forms which can give semi-quantitative answers, not only to the above problems, but to those provided by radio galaxies and quasars and perhaps even by the universe itself.

12.1 Galactic Forms

Optical galaxies have a most notable diversity of forms, and any adequate theory of their evolution from intergalactic gas must first explain these differences. These galactic forms have been classified by Hubble (1958) and in more detail by de Vaucouleurs and other workers. In Hubble's well-known tuning-fork diagram a series of elliptical galaxies of increasing ellipticity lead to an SO galaxy (a flat featureless disk) and thence to the two branches of spiral galaxies, some with circular and some with bar-shaped central regions. There is a suggestion of a possible evolutionary sequence in this diagram, but this has no basis in observational evidence. In Figure 12.2 we have sketched in plan and in side view what we consider the three basic forms of galaxy. This simpler classification neglects the distinction between normal and barred spirals and, more important, it sets ellipticals (**c**) and SO galaxies (**a**) as the extreme cases, on either side of the more average spiral (**b**). The reason for this will appear shortly.

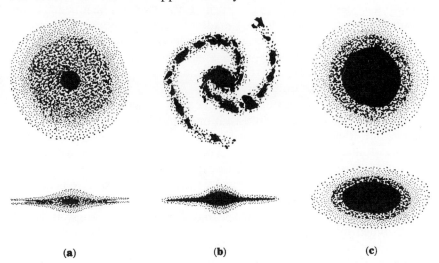

(**a**)   (**b**)   (**c**)

Fig. 12.2. The three basic forms of galaxy shown schematically in plan (along the rotational axis, top) and in side view. (**a**) The SO galaxy, one extreme form. (**b**) The spiral (barred spiral not shown), considered the "average." (**c**) The E (elliptical) galaxy, at the other extreme.

The elliptical (E) galaxies are spheroids with axial ratios varying from unity (EO) down to about 0.3. The thin Es bear a superficial resemblance to SO galaxies and at first they were related in an assumed evolutionary sequence. However, there are important differences in both morphology and behavior. The E galaxies have isophotes which are ellipses, no matter how thin, as shown in Figure 12.2c; the SOs have isophotes which are extended in the equatorial plane, being pulled out into flat disks (Figure 12.2a). The SOs also occur in unusually rich clusters and never show activity whereas the Es do not cluster as notably but grow to larger proportions and sometimes become radio galaxies, following an explosion.

SO galaxies, like spirals, are flat disks but then differ unexpectedly in having no spiral structure, no gas and no young stars. The difference was once attributed to collisions, but Baade has found many field SOs which could not have collided. Another alternative explanation of the various forms of galaxy was that one form evolved from another, but this was disproved by Baade (1963), who observed very old stars in all nearby galaxies regardless of form. Since the stars are old, all of the galaxies must be old and this rules out evolution between forms.

Apparently each galaxy has an evolutionary blueprint provided when it is born and follows this blueprint until its death. The problem is the nature of the blueprint, and is most challenging in the case of spiral galaxies which provide such a variety of phenomena. They comprise a pair of spiral arms made of new, luminous stars and gas clouds, and a central system which is often the seat of activity. Sometimes the central system explodes and sometimes, perhaps as a result of activity in the central system, a radio corona of magnetic field and cosmic-ray electrons develops.

Various theories have been developed attempting to explain the galactic forms in terms of gravitational and hydrodynamic forces. In particular, there are the theories of formation of spiral arms which are mentioned in the following section. Generally gravitational theories concentrate on one or two features to the complete neglect of others, such as the origin of radio galaxies or the reason for explosions in spiral systems or the existence of optical "bridges" between galaxies. There is now a growing body of evidence that galactic morphology is largely determined by magnetic fields and that the blueprint which determines the form that a galaxy will develop is magnetic in character.

Dynamic effects of a galactic magnetic field were considered by Chandrasekhar and Fermi (1953), in connection with the stability of spiral arms and the motions of gas clouds. The origin of the galactic magnetic field by the action of differential rotation on a very weak radial field was proposed by Piddington (1958). A more positive step towards a hydromagnetic theory of galactic structure was made by Hoyle and Ireland (1960) who tried, unsuccessfully, to account for spiral arms in terms of magnetic stresses. So far the ideas centered mainly on spiral arms and spiral fields, but Vorontsov-Vel'yaminov (1964) drew attention to other structural features suggestive of the influence of magnetic fields. These include loops, figure eights and gamma forms which are reminiscent of magnetic phenomena seen in the solar atmosphere and in the laboratory. Another phenomenon which might be attributed to magnetic forces is deformation of the gas or dust disks; Zasov (1965) found that galaxies with a curved dust layer usually belong to close groups or small clusters. Finally, Gershberg (1965) explained the intergalactic bridges of matter in terms of containment by a magnetic field.

The marked differences in the forms of galaxies, their complex structural features and the occasional eruption of a normal galaxy into a radio galaxy (Section 12.3) seem impossible to explain without invoking magnetic forces. Such considerations led to the formulation of a unifying theory in which the different forms and other characteristics are all accounted for by a single parameter (Piddington, 1964; 1967).

This theory starts with an intergalactic field whose strength may be as weak as 10^{-9} gauss and which is fairly uniform over distances of 300 kpc or more. The single parameter which determines the form of the galaxy is the angle β, between the intergalactic field $\mathbf{B_0}$ and the rotational vector ω_0 of the intergalactic plasma which forms the galaxy. The gas density is taken as that of the local cluster ($\sim 5 \times 10^{-25}$ g/cm^3) from which gas, a cloud of dimensions about 300 kpc and mass 2×10^{44} g or 10^{11} M$_\odot$ contracts to form the pregalaxy. The field is frozen into the gas and after contraction the cloud and field may take various forms, depending on the angle β. The two extreme cases $\beta = 0$ (or π) and $\beta = \pi/2$ (or $3\pi/2$) are shown schematically in Figure 12.3; there will be, of course, intermediate configurations with all values of β.

In Figure 12.3a the differential rotation of the gas cloud does not bend or stretch the field lines because the latter lie in the shear planes. The lines within the galaxy are in isorotation and a helical hydromagnetic twist wave radiates away from each side of the disk. The plasma may flow into a sheet whose thickness is determined only by random motions and not by magnetic pressure. We believe that this magnetic-rotational configuration provides an SO galaxy which, although topographically similar to spirals, has no gas or arms. As seen in Section 12.2, arms and other galactic structure may be accounted for in terms of magnetic-field deformation and consequent stress. These

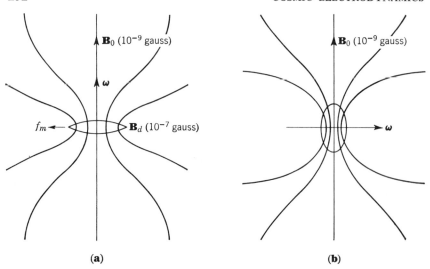

(a)   **(b)**

Fig. 12.3. A schematic representation of the extreme forms of protogalaxies and their magnetic fields before winding. (a) The intergalactic field $\mathbf{B}_0$ and rotational vector are parallel. There is no field winding within the galaxy and an SO system develops. (b) The two vectors are orthogonal and an E galaxy develops, perhaps later becoming a radio galaxy.

are lacking in SOs and hence structure within the disk is lacking. Also, since contraction of the gas cloud is not impeded by any magnetic or centrifugal force, both galaxies and stars within these galaxies may form more easily. This accounts for the rich clusters of SO galaxies and the lack of gas and dust. Finally, it is evident from the shape of the magnetic field lines that a magnetic force, f_m will tend to pull the gas into a disk of larger diameter and so account for the difference between the isophotes of E and SO galaxies which otherwise are very similar. Thus the unifying hypothesis accounts for all of the main features of SO galaxies.

The configuration of Figure 12.3b is thought to lead to the development of elliptical and radio galaxies and perhaps also of quasi-stellar radio sources or quasars. The self-gravitation of the plasma cloud is opposed in the plane of rotation by centrifugal forces and so provides the usual disk. In addition, it is opposed in the perpendicular directions by magnetic pressure and this is thought to slow contraction until stars are formed. In this way a spheroidal star system or E galaxy is formed, the ellipticity of the isophotes depending on the magnetic field strength as well as cloud density and perhaps other factors. The winding up of magnetic field lines in this configuration and the consequences are discussed in Section 12.3 and in Chapter 13.

Intermediate between the two cases shown in Figure 12.3 are a full range of oblique fields which are thought to give rise to spiral galaxies. Here the unifying theory invokes the sharp gradients in gas density and other parameters perpendicular to the disk, together with an external hydromagnetic pressure which varies with azimuth and which controls the rate of star formation and forms a standing wave of young stars. The theory has not differentiated between the normal spirals and barred spirals but below the hypothesis is advanced that the difference depends largely on the strength of the oblique field. If strong enough, the cloud collapses into a more or less uniform disk, but if weaker than a critical value, there is Coriolis acceleration which forms the disk into a bar where it is held permanently by non-radial gravitational forces.

An alternative hydromagnetic theory of some galactic forms has been developed in less detail by Pikel'ner (1965).

The hydromagnetic wave theory of spiral galaxies is given in the following section, but first we should anticipate a criticism which has been made of all "open"-type galactic magnetic field models. This is that cosmic rays would escape too readily and make it too difficult to account for the observed cosmic-ray flux. The answer is that although we start with an open-field model, differential rotation and other plasma motions eventually change this to a closed-field model.

12.2   Spiral Galaxies

The spiral systems of stars are the most spectacular of the optical galaxies, as evidenced by the beautiful spiral M51 reproduced in Figure 12.1. Theories of such systems are complex and some have been developed without considering all-important observational features. Hence, prior to discussing theories, we list what we consider the most significant features; to these must be added the fact that other quite different galactic forms have been created simultaneously and presumably from similar intergalactic gas.

(1) Spiral arms are often irregular, consisting of short sections often projecting as spurs. However, in many we see a single pair of arms starting from diametrically opposite points in the central system and extending in a continuous pattern to the boundary of the disk and sometimes even as a bridge to a nearby galaxy, as in M51 (Figure 12.1), NGC 7753, NGC 3808, and others. One must infer the existence of a non-axisymmetric field of force extending beyond the galaxy and not sharing the galactic rotation.

(2) A second major feature of spirals concerns the nature of the

arms: they comprise very young, bright stars and the gas from which these stars formed (Sandage, 1961) less than 10^8 years ago. Although the arms are striking optical objects, they represent only a small proportion of galactic mass (for M31 about 15%) so that Mayall has likened them to frosting while older stars make up the cake. One cannot escape the inference that spiral arms are quasi-permanent waves of star formation with a period equal to one-half the period of rotation of the galaxy; they must be standing waves.

(3) The third major feature of spirals is the central spheroidal system, which has anomalously high density and low angular momentum per unit mass (Baade, 1963). It is often the site of great explosions and sometimes throws off rings as shown in Figure 11.2; it often becomes a bright radio source, showing the presence of enhanced amounts of magnetic flux and cosmic rays. It is difficult to avoid the inference that galactic central regions, like protostars, are magnetically braked and create their own magnetic flux by differential rotation.

The basic difficulty in any theory of spiral arms is kinematical, amounting to overwinding; whatever the origin of the arm it will be quickly wound into a tight spiral unless it breaks up and reforms every revolution or so. This has led to the formulation of gravitational theories of disks, reviewed by Lin (1967). However, these theories fail to explain two of the three major features of spiral galaxies which are listed above as well as some minor features mentioned below, and also fail to account for different galactic forms or radio galaxies. The earlier attempts to formulate a magnetic theory also failed because it was assumed that the field would lie in the disk as shown in Figure 12.3b. However, there seems no reason why skew fields should be excluded; indeed these are to be anticipated as the average situation.

Let us consider the contraction of a protogalaxy from a diameter of a few hundred kiloparsecs to one-tenth that diameter and consequent strengthening of the field by a factor of about 10^2. In a cylindrical system of axes r, θ, z the rotational axis is in the direction of z and the field $\mathbf{B}$ at an angle $\beta \sim \pi/4$ to the z axis; the plane containing ω and $\mathbf{B}$ is the meridian plane $\theta = 0, \pi$ as shown in Figure 12.4a. The theory of galactic contraction is not understood and we may only discuss the effects in rather general terms. First we consider the plasma and the sections of the field lines inside the protogalaxy and neglect external particles and fields, although as seen later these become very important. Contraction of the galaxy with velocity $\mathbf{v}(r, \theta, z)$ will depend on

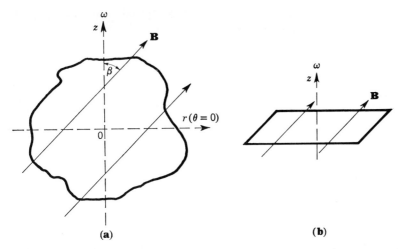

Fig. 12.4. A protogalaxy with a skew magnetic field **B**. (a) Before contraction commences. (b) Approaching galactic dimensions the plasma is constrained to contract along the field lines and to move in elliptic orbits, which differ with the distance from the equatorial plane.

gravitational force, pressure gradients, magnetic force and Coriolis acceleration $2\omega \times \mathbf{v}$. Below, we advance the hypothesis that the Coriolis force is responsible for the formation of barred spirals. Meanwhile we consider the formation of normal spirals.

Let us assume for the time being that the magnetic field inside the protogalaxy retains its general form and limits plasma motions to the cloud rotation together with motion along the field lines. The result is a disk as shown in Figure 12.4b with plasma moving in elliptical orbits, the ellipticity being given by (Piddington, 1967(a))

$$e(r,z) \sim \frac{z}{r} \tan \beta \qquad (12.1)$$

In our galaxy a typical value of e would be 0.01, and the variation of v through the disk thickness would be only 4 km/sec, which seems in agreement with observed differences. This is a case of modified iso-rotation with the magnetic field lines lying on the plasma shear surface so that there is no winding up, even though there is a substantial field component B_r.

Now suppose that while plasma density and velocity distributions remain unchanged, the field is tilted so that the skew angle increased to $\beta + \delta$. In the sectors $\theta \sim 0, \pi$ there is now a field component B

cos δ across the shear planes and so an azimuthal component develops as a result of differential rotation.

$$B_\theta = -B\cos\delta tr \frac{\partial\omega}{\partial r} = 2AtB\cos\delta \tag{12.2}$$

Where t is the time elapsed and A is the constant of differential rotation. In our galaxy, near the sun $A \sim 15$ km/sec/kpc and so

$$B_\theta \sim 30t'B\cos\delta \text{ (t' in units of 10^9 years)} \tag{12.3}$$

In the galactic sectors $\theta \sim \pi/2$, $3\pi/2$ the additional tilt of the field (δ) is parallel to the shear surface and so no winding up occurs. Thus, initially we have four sectors in two of which shear will occur and B_θ will grow, and in two of which no shear will occur and no B_θ will develop. As winding proceeds these sectors themselves are wound into four separate spirals, each of about twenty turns. In our own galaxy, the sun is in a sector-spiral in which field winding has occurred (as evidenced by the observed field).

So far we have a disk magnetic field which is either tightly wound or not wound at all, but no suggestion of a pair of open spirals as required. These comprise gas clouds and stars formed from these clouds, and these clouds must form from the more tenuous gas as a result of velocity gradients. The ratio of the diameter to thickness of the disk of our galaxy and many others is large, about 100 or more, and it is clear that gas motions perpendicular to the disk will be much more effective than similar motions parallel to the plane of the disk. It is surprising, therefore, that most spiral arm theories concentrate on the latter. In the unifying theory, motions perpendicular to the disk are held responsible for spiral arms, and these are caused by varying external magnetic forces.

In Figure 12.5 a pinched hour-glass galactic magnetic field is shown, formed by the contraction of the gas cloud with its skew field shown in Figure 12.4; rotational effects are not shown. The field strength in the disk is initially 10^{-7}–10^{-6} gauss but due to the development of an azimuthal field in some sectors and the compression of the remaining field after expansion of these sectors, a strength of a few times 10^{-6} gauss develops. One effect of these fields which is immediately obvious is the mechanical force shown as **F** which bends one edge upwards and the opposite edge downwards. The above field strengths are adequate, and observational evidence of such distortion is found in our galaxy, in M31 and in many other systems. The distortion in the plane of our galaxy has been explained in terms of the gravitational field

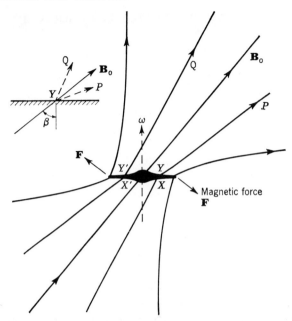

Fig. 12.5. A schematic representation of the oblique magnetic field model of a spiral galaxy. The tilted hour-glass field bends the edge of the disk down at the right and up at the left. Similar asymmetric magnetic forces perpendicular to the disk may give rise to patterns of accelerated star formation and hence to spiral arms.

of the Magellanic Clouds, but even if this is the correct explanation it does not apply for isolated galaxies.

Galactic rotation will twist the external field into a helical form as shown in Section 4.1, and a twist wave will radiate away from both sides of the galaxy at a velocity which approximates the Alfvén velocity. The result, after a long period, is a standing hydromagnetic wave and an external disk pressure which varies with azimuth. Consider a field intercept XY connecting to field line P which, after a half revolution, becomes $X'Y'$ connected to line Q. The two surface field configurations are shown in the insert and since the component B_z must be continuous through the surfaces of the disk, the external field near Y is stronger than that near Y'. The pressure maximum and minimum on the top of the disk are originally in the segments $\theta \sim 0, \pi$, but these segments are twisted into a spiral form. A second spiral of external pressure maximum, displaced by one half revolution, is formed on the bottom of the disk. Magnetic field strength changes from about 1 to 3×10^{-6} gauss are enough to vary the disk thickness by a factor of about 2 and so, presumably, to profoundly change the rate of star formation.

The unifying theory thus provides a semi-quantitative explanation of the origin of a pair of diametrically opposite spiral arms fixed in relation to nearby galaxies and perhaps connected to one of them by a bridge of field and gas.

We may be permitted to speculate on the modifications of this model needed to produce a barred spiral galaxy.

An essential feature of the normal spiral model is that during collapse from cloud to disk (Figure 12.4), circular symmetry is preserved. It will be seen that gas moving along the field lines with velocity $\mathbf{v}$ will tend to develop a Coriolis acceleration $2\omega \times \mathbf{v}$. This tends to move the gas perpendicular to the field and would result in distortion and stress of the field, but we have assumed that $\mathbf{v}$ is small enough and $\mathbf{B}$ is large enough to effectively eliminate such motion. However, if this is not the case, then in the sectors $\theta \sim 0, \pi$ gas is everywhere given an angular acceleration v_θ, while in the sectors $\theta \sim \pi/2, 3\pi/2$ there is no acceleration because motion $\mathbf{v}$ is more or less azimuthal. The result is that the whole gas cloud tends to collapse into two sectors and if this motion can continue far enough (in spite of magnetic stresses), then non-radial gravitational forces take control and the cloud collapses to a bar rather than a disk. In this way the model may account for barred spirals.

The same oblique-field model, in the case where the cloud collapses to a disk, also accounts for the main features of the central region. This development is traced in the following section.

12.3   Central Systems and Coronas

Spiral galaxies comprise two contrasting optical systems and some-times also a radio corona. In normal spirals most of the angular momentum of the protogalaxy is stored in the disk, while on the other hand the central system has acquired far too much mass. In our galaxy the average mass density of the central system is about 1000 times that near the sun. In barred spirals the central system is a solidly rotating bar often making up most of the galaxy. These distributions constitute serious theoretical problems, reminiscent in some ways of the problem of star formation. In the case of the elliptical galaxies, and in particular the spherical EO systems, there are no spiral arms and the central systems are usually very small and bright. On short exposure the "photograph" of an elliptical shown in Figure 12.2c might show a central dot, too small to be resolved and perhaps only 0.1 kpc or less in diameter.

A more remarkable feature of the central systems is the violent activity often seen there (Ambartsumian, 1960). "Explosions" occur in many galaxies, notably in Seyfert's and on vastly greater scales in elliptical galaxies which become radio galaxies. Observations of these explosions have been reviewed by Burbidge *et al.* (1963) and Pacholczyk and Weyman (1968). The results of these explosions seem to be short-lived, about 10^7 years or less, compared with the life of a galaxy of about 10^{10} years, and from this we infer that many galaxies may have exploded at one time or another and will explode again. A striking example of an exploding galaxy is the peculiar spiral M82 (Lynds and Sandage, 1963) from whose central system plasma is moving with velocity about 1000 km/sec along the rotational axis on both sides of the disk. An explosion seems to have occurred about 10^6 years ago and now plasma filaments and magnetic field lines extend outwards for about 3 kpc. The mass of ejected material is about $6 \times 10^6 M_\odot$ and kinetic energy is about 2×10^{55} erg.

An important feature of activity in galactic central regions is that these regions become strong radio sources, with luminosities in the range 10^{36} erg/sec for our own rather dim central region to nearly 10^{45} erg/sec for Cygnus A, with a full range of intermediate values. Most of this emission is synchrotron (Section 4.5) and indicates huge concentrations of cosmic-ray and magnetic energy, at least 10^{54} erg in our central region and much more in the radio galaxies. Sometimes these great concentrations of particles and fields in spiral galaxies burst out to form a corona (a term preferred to halo, which refers to a stellar population). M82 has such a corona and others have been detected by their radio synchrotron emission.

An adequate theory of the central systems and coronas of spiral galaxies may be a start towards a theory of quasars and radio galaxies. For this reason we should bear in mind the energy requirements of the latter, which may amount to 10^{61} erg or more. The energy available if a whole galaxy could be converted from hydrogen to helium is only about 10^{63} erg and this fact prompted Hoyle and Fowler to suggest gravitational energy as an alternative. The difficulty which is often overlooked, however, is that a specific mechanism is required to convert part of the gravitational energy to magnetic energy and part to cosmic-ray energy. For a review of models which neglect these conversion mechanisms see Maran and Cameron (1964).

Attempts to link gravitational collapse to a magnetic field system have been made by Piddington (1964; 1967), Kardashev (1964), Ginzburg and Ozernoi (1964), Pikel'ner (1965), Ozernoi (1966) and Sturrock (1965, 1966) (see also Chapter 13). In choosing between theories many factors should be considered. These include the ability of the model to eject two ordered magnetic-field systems along the rotational axis of the parent body, and to repeat this process many times; to operate in the central systems of spiral galaxies as well as in radio galaxies; to "harness" gravitational energy so that it may be abstracted steadily between explosions and, finally, to operate on a time scale as short as one year. The spiral galaxy model discussed above seems to provide a beginning for a model of exploding central systems and may be extended to quasars and radio galaxies.

Winding up of the magnetic field in the disk of a galaxy may be avoided if the plasma is forced into elliptical orbits given by equation 12.1 and illustrated in Figure 12.4. However, within the central region where r is small and a field line may even cross the axis of rotation, winding is inevitable and its results have been discussed in detail by Wentzel (1960) and Ôki et al. (1964). The field component B_s across the shear surfaces results in the development of an azimuthal component according to equation 12.2, and together these provide a magnetic force of density F_θ given by

$$4\pi F_\theta = \frac{B_s}{r} \frac{\partial}{\partial r} (rB_\theta) \qquad (12.4)$$

This force may accelerate or decelerate the gas; when rB_θ decreases with r, acceleration occurs. Near the center rB_θ must be increasing from zero and braking occurs. Far enough out B_θ must vanish and within the region where rB_θ is decreasing, acceleration occurs. Putting in reasonable values of field strength and gas density we find a time scale of about 10^{10} years and so the force F_θ, while too small to cause spiral structure, may account for the transfer of most angular momentum from the central region to the disk.

Within the central region, where braking occurs, the loss of angular momentum will cause the plasma to move to lower Keplerian orbits. There will be a general shrinkage and release of gravitational energy. Outside the central region the disk expands and the galaxy assumes the form shown schematically in Figure 12.6. The dotted region is the disk, which may have its own azimuthal field B_d (if there has been winding) and in any case is threaded by the original compressed intergalactic field B_o. The disk is connected to the central region C by a spiral field B_g and within C there may be a solid rotating nucleus N. By redistribution of the magnetic stresses within C it is possible for other arms or rings to be spun off, just as the 3 kpc arm is moving away from the central system of our galaxy (Figure 11.2).

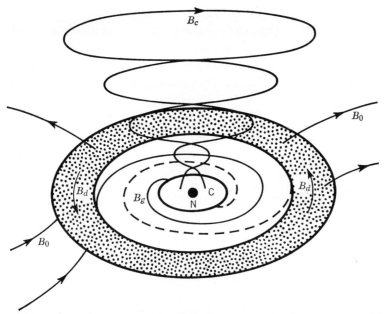

Fig. 12.6. A highly schematic representation of an evolving spiral galaxy. The original oblique field B_0 is wound into a spiral which forces the disk component (dotted) outwards and the central spheroidal system C inwards. An outburst may occur between a "solid rotation" nucleus N and the outer boundary of C, and this creates a radio corona with field B_c.

We will consider briefly how gravitational energy is converted steadily to other forms. As seen from equation 12.2, the azimuthal field B_θ grows continuously. A typical differential rotation rate in central systems is 100 km/sec/kpc and in the lifetime of a galaxy an initial radial field of only 10^{-6} gauss will develop an azimuthal field of 10^{-4}–10^{-3} gauss. During the period of braking and slow loss of angular momentum, the plasma moves in Keplerian orbits with velocity V_θ. If the characteristic mass and radius of the central system are M and R, then the balance of gravitational and centrifugal forces requires

$$V_\theta^2 = GMR^{-1} \qquad (12.5)$$

where G is the gravitational constant. The total gravitational and magnetic energies of the cloud are

$$\Omega \sim GM^2R^{-1} \qquad W_m \sim \frac{1}{6}B^2R^3 \qquad (12.6)$$

so that $\Omega \sim 2K$, where K is the rotational kinetic energy. Thus as the cloud shrinks, half the gravitational energy is converted to kinetic energy, and half to magnetic energy (neglecting other energy sinks). If we take $M = 10^9$ $M_\odot$ (10% of the total mass of our central region, now mainly in the form of stars) and allow this massive cloud to shrink to a radius of say 300 pc, then the kinetic and magnetic energies are each about 10^{56} erg. This is the energy requirement for a small to average radio galaxy.

This orderly growth of a toroidal magnetic field must eventually be interrupted by the onset of two new phenomena, which we consider briefly.

(1) In the spiral neutral sheet between oppositely directed magnetic field sectors (illustrated in Figure 4.1), magnetic field annihilation will occur with conversion of energy to particle kinetic energy including, perhaps, cosmic-ray energy. This process may explain the development of radio brightness in the central regions of some spiral galaxies and also, perhaps, the primary cosmic radiation in our galaxy (Section 12.5).

(2) The second effect which must eventually interrupt the field winding process is the onset of Rayleigh-Taylor instability and the bursting of loops of magnetic field through the two sides of the galactic central region. This phenomenon has been described in Section 11.1 and is illustrated in Figure 11.1. Now it is identified with the formation of galactic coronas, illustrated by the magnetic field B_c of Figure 12.6 (the field below the disk is not drawn). It is also identified with the ejection of plasma from the central regions of galaxies, such as M82. In the central region described above, kinetic and hydromagnetic velocities are typically 100 km/sec and the mass is 10^9 $M_\odot$. If 1% of this mass is ejected by the Rayleigh-Taylor instability, then the velocity of ejection should be about 1000 km/sec. This is the mass and velocity of material ejected from M82.

Thus the spiral hydromagnetic model of galactic central regions seems to provide a semi-quantitative explanation of the observed high density with low angular momentum, unexpected structural forms, high cosmic-ray density, and outbursts of plasma and magnetic field. In the following section it is shown that it may also explain the existence of a very luminous nucleus.

12.4   Seyfert Galaxies and Related Objects

Even before the discovery of radio galaxies, Seyfert (1943) had directed attention to the central systems of some spiral galaxies which showed extremely small luminous nuclei and very broad emission lines. Little attention was paid to these objects for two decades, until the problem of radio galaxies brought energetic events in galaxies to the attention of astronomers. The study of Seyferts was given further impetus by the discovery of quasi-stellar objects, and a possible relationship based on spectral similarities was suggested by Burbidge *et al.* (1963). Other "related objects" may be the compact so-called N galaxies, which also emit strongly in the radio spectrum. The notable features of all of these galaxies are, first, a nucleus whose continuum emission in the infrared may amount to 10^{60} erg or more over a lifetime of 10^9 years and, second, broad emission lines indicating Doppler velocities up to $6-8 \times 10^3$ km/sec. Other characteristics are described by contributors in Pacholczyk and Weyman (1968).

The new problem posed by Seyfert, and other, nuclei is the continuum emission amounting to $10^{60}-10^{62}$ erg over a period of perhaps 10^9 years. Such total energy is comparable with that for radio galaxies and, as we shall see in the following chapter, seems explicable only in terms of gravitational energy. The gravitational energy available from a given mass distributed over a given radius may be found from equation 12.6, and the requirement is satisfied by a mass of $10^{10} M_\odot$ and radius about 0.1–10 pc. However, such an object cannot be supported by gas pressure as in ordinary stars and tends to collapse gravitationally and give up its energy within a period of order 10^7 years or less instead of 10^9 years, as required.

Several models devised to overcome this difficulty are discussed in Pacholczyk and Weyman (1968). Colgate envisages a self-gravitating cluster of stars in which occur multiple stellar coalescences followed by rapid supernova events. This and other similar models may meet the requirement of 10^{60} erg or more delivered over a period of about 10^9 years and so explain the optical nucleus. However, the models are ad hoc and neither their origins nor their relationship to other phenomena are evident. Sturrock and Feldman envisage an intergalactic magnetic field linked through the compact object; the field removes angular momentum from incoming intergalactic gas and allows it to be accreted and so to provide energy. This model is an extension of a radio-galaxy model discussed in the following chapter. It does not relate to other features of spiral galaxies (forms and structural features of central regions) and as Woltjer has pointed out it is difficult to satisfy the energy requirement by accretion.

While many models of Seyfert nuclei are ad hoc and unrelated to other problems, a simple extension of the magnetic spiral field hypothesis seems to provide a more justifiable model. Here we have a central plasma cloud which spins, winds up its magnetic field and so is braked and contracts gradually. Thus we have a simple explanation of the origin of a system of such high density, of its failure to collapse under gravitational force, and perhaps its failure to fragment, this being inhibited by the strong magnetic and inertial force fields. As we have seen, this model may explain the deficit of angular momentum observed in the central systems of spirals and also their high density. A logical extension is for a small nuclear region to be braked virtually to a standstill and so to commence a Helmholtz contraction as in the case of a protostar (Section 11.2). This is a superstar whose creation was made possible by the stabilizing influence of the magnetic field and the spin.

The braking force density is given by equation 12.4 and when this is related to the angular momentum an order of magnitude time of decay of the latter is given by

$$T \sim MV_\theta R^{-2} B_\theta^{-1} B_s^{-1} \qquad (12.7)$$

Using the above values of B_θ, B_s, R, V_θ and M we find a characteristic time of a few times 10^9 years. All five variables on the right-hand side of equation 12.7 are functions of R, and the variety of models is endless. We envisage the creation of a relatively small nucleus which begins to collapse as a superstar. There must be a continuous flow of angular momentum outwards from the whole central region and so successive shells will then be braked sufficiently to collapse onto the superstar and give up their gravitational energy. The time scale for the collapse of the whole central region will depend on the various parameters of equation 12.7 and may have a value of about 10^9 years, which fits the observational data.

Thus, according to the spiral field hypothesis, the winding up process may lead to one or other of two catastrophes: an outburst to provide a magnetic corona, or a superstar to provide a Seyfert-type nucleus.

12.5   The Origin of Cosmic Rays

Cosmic rays provide us with our only direct sample of matter originating outside the solar system. They play important parts in the dynamics of interstellar gas clouds, of the galaxy as a whole and perhaps of the intergalactic medium. Their radio synchrotron emission reveals numerous concentrations of magnetic fields and cosmic-ray electrons throughout the universe, and these in turn must be interpreted in terms of a variety of electrodynamic phenomena. The origin of the primary cosmic radiation is not known, but acceleration must, presumably, be by one of the mechanisms described in Section 4.4. From observations of solar cosmic rays and Van Allen particles it would seem that magnetic neutral sheets provide the best basis for a model, although Fermi acceleration may be important, particularly in providing the most energetic particles.

Direct measurements show an energy density of primary cosmic radiation near the solar system of about 10^{-12} erg/cm^3, which may be compared with that in a quasar of perhaps 0.1 erg/cm^3 or more. The local cosmic rays are mainly protons with small amounts of heavier nucleons in proportions which suggest a thermonuclear origin (Rees and Sargent, 1968). There is also an electronic component amounting to only about 1% of the proton energy, but of great significance because it links the local observations with radio synchrotron data. These data are disparate in that only the electron component is observable and then only when moving in a magnetic field of sufficient strength. Extragalactic sources comprise radio galaxies and quasars (Chapter 13) as well as many spirals exhibiting central activity (Tovmasyan, 1967). Galactic synchrotron sources are mostly supernova remnants, but include the central region and possibly a corona. The sun (Section 5.6) and Jupiter (Section 10.2) are also synchrotron sources. The problem is to relate these two sets of data and decide where the primary cosmic radiation originated. The most popular theories of origin are in galactic objects such as supernovae or in extragalactic "violent events."

The total cosmic-ray energy in the galaxy is estimated as 10^{54} erg and it seems that this must be replenished every 10^6 years or so, otherwise the heavier nucleons would be broken up and the relative numbers of these and protons would not be as observed. Most of the regions in the galaxy where intense fluxes of relativistic electrons are known to exist are near supernova remnants and so these have been investigated as the main source of cosmic radiation. A consistent theory has been developed by Ginzburg and Syrovatskii (1964). There may be a supernova every few hundred years and if each provided 10^{51} erg of cosmic-ray energy, then the requirement would be met. The Crab Nebula is providing cosmic-ray electrons at a rate of about 10^{49} erg/10^3 years. It may have a lifetime several times 10^3 years; it may have provided at a higher rate in the past, and it may provide more cosmic-ray protons than electrons. Thus if the Crab is a typical supernova remnant, this source of cosmic rays seems adequate.

There are a number of difficulties met by the supernova theory. Even allowing for its youth compared with other supernova remnants, the Crab may be a unique object, producing at a much faster rate than others at the same age. It may not produce more fast protons than electrons, and all fast particles produced must undergo severe betatron deceleration in the outwardly expanding magnetic field (Section 11.5).

Gold has suggested that the Crab energy source is a rotating neutron star (Section 11.6) and that its earlier output may have been 10^3 times that at present. However, such an enormous output of cosmic-ray electrons should have been accompanied by a corresponding increase in optical radiation which should have been detected. Finally, there is the difficulty that cosmic rays generated in a cloud of dimensions very small compared with the galactic dimensions are likely to escape rapidly from the galaxy as a bubble of gas escapes from a denser liquid.

The alternative, widely discussed theory of the source of primary radiation is in radio galaxies which, although much more distant, provide some 10^{10} times more cosmic-ray energy than do supernovae. The total energy requirement is 10^6 times larger because galaxies, even when given a corona of dimensions equal to the disk diameter, occupy only 10^{-6} of all space. This may not be an objection but there are others which appear fatal. If we take 10^{58} erg as the electron cosmic-ray energy of an average radio galaxy, 10^2 times more as proton energy and a total of 10^5 existing sources, then we have a total energy of 10^{65} erg in a volume of 10^{84} cm^3, or 10^{-19} erg/cm^3. On the basis that a source has a lifetime as a radio galaxy of only about 10^6 years, compared with 10^{10} years for the galaxies, this is increased by a factor of 10^4 to 10^{-15} erg/cm^3. In spite of the rather optimistic values used, the theory still fails by a factor of 10^3. Furthermore, recent work described in Chapter 13 shows that radio galaxies do not die merely by dispersion but that the cosmic rays suffer severe betatron deceleration and so even less cosmic-ray energy is finally dispersed.

The theory may be modified to place the source of cosmic radiation within the Virgo supercluster (volume 10^{76} cm^3, containing several radio galaxies), or within the local group of galaxies (volume about 10^{72} cm^3, no radio galaxies), but these variations seem to meet similar difficulties.

The above two theories are based largely on radio observations of synchrotron emission from two classes of sources. A third class of source seems to us to offer a better chance of explaining the primary cosmic radiation; this is the galactic central region. Radio observations of spiral galaxies sometimes show a component of uniform brightness over the disk, together with a small central component of comparable luminosity (Mills and Glanfield, 1965 and references given there). In our own galaxy the small central region has a luminosity about 1% of the whole. If we accept the low estimate of the galactic magnetic field (Section 11.1), then in order to account for the galactic synchrotron emission the cosmic-ray electron flux must increase considerably in the general direction of the central region. This provides some evidence that the central region is a major source of the local flux.

Further evidence is provided by the central regions of other spiral systems. Not only do some show very much larger concentrations of cosmic rays, far more than enough to fill our galaxy, but they show that these concentrations may appear within a short period. In M82 the time scale of activity, including the appearance of cosmic rays, is only about 10^6 years. We do not suggest that any such catastrophic event has occurred in our galaxy, at least for a very long time, but that cosmic rays are being created in the central region at a rate exceeding 10^{48} erg/year. Most of these escape so that only a quantity corresponding to 10^4 years' production is stored there at present. Unlike the other models, this puts no great strain on the source because the central region need only produce at a rate about 100 times that of the Crab Nebula. The central regions of some other spirals produce at much greater rates, and those of developing radio galaxies at rates as much as 10^9 times greater.

The difficulty met by this model is in the movement of cosmic rays out to the sun. Diffusion through the disk with its wound up magnetic field is likely to be too slow and the alternative path through the corona seems more likely. Here we have a problem in containment: the magnetic field of the corona tends to be disrupted by the cosmic-ray gas and so a steady-state model is not valid. However, there is some evidence that the corona is not in a steady state; for example, clouds of atomic hydrogen are seen falling into the disk (Section 11.4). Perhaps the model shown in Figure 12.6 may be extended to include cosmic-ray transfer.

Within the central region of our galaxy the magnetic field is being wound into a tighter and tighter spiral. Periodically the magnetic pressure and cosmic-ray pressure must build up until the Rayleigh-Taylor instability causes magnetic loops and cosmic-ray gas to burst out in the manner illustrated in Figure 11.1 The result is a coronal field something like B_c of Figure 12.6, which is rooted in the central region and extends perhaps 15 kpc. If the field strength in the corona is 10^{-6} gauss, then the magnetic flux is a few times 10^{39} gauss cm^2 and the mass required to anchor this flux by gravitational force within the central region is about 10^9 M$_\odot$, which is reasonable. The coronal field and cosmic-ray gas both tend to expand and they will press against the galactic disk and allow a flow of cosmic rays into the disk. This flow might be aided by the neutral hydrogen gas clouds ejected during the outburst and falling into the disk in the region of the sun.

The galactic spiral-field model also includes a magnetic neutral sheet, illustrated in Figure 4.1, and this is a likely region of cosmic-ray acceleration. The acceleration mechanism is not fully understood and the relative numbers of cosmic-ray protons and electrons is not known.

However, there is evidence that in solar neutral sheets electrons may sometimes be accelerated with the same efficiency as protons, but may not escape as easily. Perhaps this is also the case in the galaxy and if so, the electron component is relatively larger towards the central region. Another mechanism for accelerating cosmic rays is the *magnetic rotor*, whose various forms are discussed in Section 11.6 in connection with pulsars. A rigidly rotating galactic nucleus with its frozen-in poloidal field is a magnetic rotor of mass perhaps 10^8 times or more that of a neutron star. Such objects may be the basic source of cosmic rays.

Support for the view that cosmic-ray electrons are chiefly primaries (rather than the results of collisions with heavy particles) is found in recent measurements of the relative proportions of electrons and positrons (Hartman, 1967). Theory shows that collisions would provide an excess of the latter, while above 1 BeV the measurements show less than 10% positives. As seen in the following chapter, a primary source of cosmic-ray electrons reduces the difficulties met in explaining radio galaxies and quasars where it has been customary to assume proton energies 100 times that of the electron cosmic rays needed to explain the observed synchrotron radiation.

This result may not apply to the cosmic-ray electrons of much higher energies. There is some evidence that positrons predominate above 10 BeV—at least near the earth.

References

Ambartsumian, V. A., 1960, *Observatory* **80**, 125.

Baade, W., 1963, *Evolution of Stars and Galaxies,* Harvard University Press, Cambridge.

Burbidge, G. R., Burbidge, E. M. and Sandage, A. R., 1963, *Rev. Mod. Phys.* **35**, 947.

Chandrasekhar, S. and Fermi, E., 1953., *Astrophys. J.* **118**, 116.

Gershberg, R. E., 1965, *Soviet Astron. AJ* **9**, 259; *Astron, Zh.* **42**, 330.

Ginzburg, V. L. and Ozernoi, L. M., 1964, *Soviet Phys. JETP* **20**, 689; *Zh. Expt. Theor. Phys. (USSR)* **47**, 1030.

Ginzburg, V. L. and Syrovatskii, S. I., 1964, *The Origin of Cosmic Rays,* Pergamon Press, Oxford.

Hartman, R. C., 1967, *Astrophys. J.* **150**, 371.

Hoyle, F. and Ireland, J. G., 1960, *Mon. Not. Roy. Astron. Soc.* **120**, 173.

Hubble, E., 1958, *The Realm of the Nebulae,* Dover, New York.

Kardashev, N. S., 1964, *Soviet Astron. AJ* **8**, 643; *Astron. Zh.* **41**, 807.

Lin, C. C., 1967, *Ann. Rev. Astron. Astrophys.* **5**, 453.

Lynds, C. R. and Sandage, A. R., 1963, *Astrophys. J.* **137**, 1005.

Maran, S. P. and Cameron, A. G. W., eds., 1964, *Physics of Non-Thermal Sources, NASA SP-46.*

Mills, B. Y. and Glanfield, J. R., 1965, *Nature (London)* **208**, 10.

Ôki, T., Fujimoto, M. and Hitotuyanagi, Z., 1964, *Prog. Theor. Phys. (Japan) Suppl.* **31**, 77.

Ozernoi, L. M., 1966, *Soviet Astron. AJ* **10**, 241; *Astron. Zh.* **43**, 300.

Pacholczyk, A. G. and Weyman, R. J., 1968, *Proc. Conf. Seyfert Galaxies and Related Objects,* Steward Observatory.

Piddington, J. H., 1958, *Proc. Inst. Rad. Eng. (U.S.A.)* **46**, 349.

Piddington, J. H., 1964, *Mon. Not. Roy. Astron. Soc.* **128**, 345.

Piddington, J. H., 1967, (a) *Mon. Not. Roy. Astron. Soc.* **136**, 165; (b) *Planet. Space Sci.* **15**, 1625.

Pikel'ner, S. B., 1965, *Soviet Astron. AJ* **9**, 1; *Astron. Zh.* **42**, 3.

Rees, M. J. and Sargent, W. L. W., 1968, *Nature (London)* **219**, 1005.

Sandage, A. R., 1961, *The Hubble Atlas of Galaxies,* Carnegie Institute, Washington Publn. 618.

Seyfert, C. K., 1943, *Astrophys. J.* **97**, 28.

Sharpless, S., 1965, *Stars and Stellar Systems V,* A. Blaauw and H. Schmidt, eds., Chicago University Press, Chicago.

Sturrock, P. A., 1965, *Nature (London)* **205**, 861.

Sturrock, P. A., 1966, *Nature (London)* **211**, 697.

Tovmasyan, G. M., 1967, *Astrofizika* **3**, 555.

Vorontsov-Vel'yaminov, B. A., 1964, *Astron. Zh.* **41**, 814.

Wentzel, D. G., 1960, *Bull. Astr. Inst. Netherlands* **15**, 103.

Zasov, A. V., 1965, *Soviet Astron. AJ* **9**, 738; *Astron. Zh.* **42**, 959.

CHAPTER 13

Radio Galaxies, Quasars and the Universe

The identification by Baade and Minkowski of a distant galaxy with the radio source Cygnus A provided one of the outstanding problems in astrophysics. The great distance indicated a radio power output of about 6×10^{44} erg/sec, which is more than the total radiation of an average galaxy. When interpreted as synchrotron emission by cosmic-ray electrons gyrating in a magnetic field, it revealed a concentration of magnetic and cosmic-ray energies in excess of 10^{60} erg. This exceeds by an order of magnitude the kinetic energy of a large galaxy moving with a speed of 300 km/sec, and on the basis of such energy considerations it would seem to provide the biggest problem in cosmic electrodynamics.

Little headway had been made with this problem when an even more remarkable object was discovered, the quasi-stellar radio source, or quasar. Many radio sources could not be identified with any optical object and as the source positions improved in accuracy, this failure became more puzzling. Finally Matthews and Sandage made an identification with a faint starlike object with broad emission lines that did not correspond with normal stellar spectra. Other identifications with starlike objects followed and in 1963, Schmidt explained their spectra in terms of very large redshifts. Within a few years, large numbers of these objects had been catalogued and then Sandage and Véron found that there is an even larger population of objects with the same optical characteristics as quasars, but with no radio emission. These have been called "blue stellar objects" and "quasi-stellar objects" or QSOs and this term is used here for the optical object with or without radio emission. Most astronomers consider the origin of the large redshifts of QSOs to have the usual cosmological interpretation, in which case some QSOs are about five times more distant than the furthest normal galaxy and bring the range of astronomical observation to the edge of the universe. Their optical emission is 100 times or more that of ordinary galaxies and so both their radio and optical outputs pose major

270

problems in astrophysics. The interpretation of the radio emissions from radio galaxies and quasars is discussed in Section 13.1, and the additional optical problem posed by QSOs is discussed in Section 13.2.

We stated in Chapter 12 that the origins of the different kinds of galaxies, including radio galaxies (and now, presumably, quasars and radio-quiet QSOs), constitutes the biggest problem in astronomy. The problem has many facets, each a part of galactic dynamics and of galactic morphology. We attempted to show in Chapter 12 that an intergalactic magnetic field must be invoked to explain galactic morphology, including the remarkable variety of galactic forms, the existence of spiral arms and other features of the disks, and a number of features of the central systems. The radio galaxies and quasars provide further, and even more remarkable, examples of activity in the central systems of galaxies and make further demands on an intergalactic magnetic field.

As we shall see below, the twin radio sources of radio galaxies and quasars appear to have emerged from the center of a galaxy, or from the vicinity of a starlike object. In fact, it was this feature of radio galaxies which first concentrated attention on the central systems of galaxies in general, and led to the discovery of the various manifestations of central activity discussed in the preceding chapter. The similarities between these manifestations have been widely discussed in the literature and have led to the idea that a single phenomenon is involved. In our view these similarities are somewhat deceiving, and there are at least three distinct, although closely related, phenomena.

(1) Hydromagnetic activity, with plasma velocities in the range a few hundred to a few thousand kilometers per second. Perhaps a prototype is found in M82 (Section 12.3), from which plasma and frozen-in magnetic field has been ejected with velocity about 1000 km/sec. In the case of spirals such as our own galaxy plasma velocities nearer 300 km/sec are observed and such outbursts may be responsible for the origin of coronas (shown schematically in Figure 12.6.). The upper limit of such hydromagnetic activity seems to occur in Seyfert galaxies, with velocities up to about 8000 km/sec and energy per gram of material about 3×10^{17} erg.

In Section 12.3 this hydromagnetic phenomenon was explained in terms of the Rayleigh-Taylor instability which must develop in the central system because any field present is wound up continuously, and increases in strength according to equation 12.3. The phenomenon is similar to the solar eruptive prominence which occurs when magnetic loops move upwards into the solar corona.

(2) Photon emission at rates between about 10^{43} erg/sec for Seyfert nuclei and 10^{46} erg/sec for QSOs. There are striking similarities between these objects, their spectra extending from the optical into the infrared and millimeter regions on the one hand and beyond the ultraviolet on the other. The sources are extremely small on a galactic scale and in each case the total energy requirement exceeds 10^{60} erg (the time scales being very different).

In Section 12.4 an explanation of Seyfert nuclei was given in terms of magnetic braking and gravitational collapse of gas shells of successively larger diameter. This model is extended in Section 13.2 to QSOs.

(3) Ejection of bubbles or streams of cosmic-ray gas from the central region of a galaxy, in either direction along the rotational axis into an existing magnetic field configuration. In Section 13.1 we consider the evidence that the radio sources leave their parent galaxy with velocities approaching that of light and energy per gram of material at least 5×10^{20} erg, and probably much more (per unit rest mass) if account is taken of relativistic effects. A clear differentiation between this phenomenon and hydromagnetic activity is made in Section 13.3, where a simple model is described.

These three different phenomena do have in common an energy requirement of 10^{60} erg (less conservative estimates range up to 10^{62} erg). It is most unlikely that more than one source of such an enormous amount of energy is available, or that this could be other than gravitational energy as suggested by Hoyle and Fowler (1963). Masses in the range 10^8–10^{10} $M_\odot$ can provide enough energy but meet major difficulties: how can such a mass contract without fragmenting, how can even a fraction of the energy be released outwards, and how can much of it be converted to magnetic energy. The inclusion of a frozen-in magnetic field seems essential in overcoming these difficulties, and although this approach is speculative it appears most promising.

In Section 13.4 we review briefly the properties of the intergalactic medium and in Section 13.5 some aspects of cosmology in which magnetic fields may play a part.

13.1 Radio Sources and their Interpretation

Of the thousands of cosmic radio sources which have been observed, most are distributed more or less uniformly over the celestial sphere and so are thought to be extragalactic. Many hundreds have been identified with particular galaxies whose redshifts range up to $\delta\lambda/\lambda \sim 0.5$. From their redshifts the source distances may be determined and these range up to several hundred megaparsecs. From the distances the radio power output may be determined, and radio luminosities are found to range from about 10^{40} erg/sec up to 10^{44} erg/sec, which exceeds the total power output of normal galaxies. These objects are the radio galaxies.

Unidentified radio sources have properties similar to those which have been identified and are assumed to be radio galaxies or quasars (Section 13.2) too distant to be photographed. However, there is a possibility of another type of object, which is discussed in the following section.

The very powerful radio sources usually appear in pairs, one on either side of the optical galaxy, lying on its rotational axis. The optical galaxies themselves may take a variety of forms, but for the most part they are the most luminous and massive of the galaxies and generally elliptical, tending to the EO spherical form. Some have great bands of dust or other peculiar features.

While these characteristics seem to define "radio galaxies" fairly definitely there are many other galaxies which are also radio sources. A number of spiral and irregular galaxies are moderately strong sources with luminosities ranging up to about 10^{38} erg/sec. However, the radio emission is not only notably weaker but it comes mainly from the central regions rather than from two external sources. These galaxies have been regarded as members of an evolutionary sequence including radio galaxies but, as seen below, it seems likely that they are quite different phenomena and only distantly related to the more powerful sources.

The structure, or radio brightness distribution, of radio galaxies has been reviewed by Moffett (1966) and many additional examples given by Macdonald et al. (1968). Typically a source has two components, one on either side of the optical galaxy, of different luminosities and different distances from the parent. Four typical radio sources are shown schematically in Figure 13.1. At the top the twin components (black) are inside the galaxy as in the case of 3C295. Below this is a source like 3C319 and Cyg A where the components are clear of the galaxy and show some elongation in the direction of the axis. In the two lower cases the sizes of the components are very different and illustrate two possible trends, either to large components as in Herc A or small as in 3C33. There are some larger than Herc A and at least one which appears to be smaller than 3C33, having a ratio of displacement to component diameter of more than 200. The average for this ratio is about 4. Other measurements of great importance in connection with source models are those of polarization of the radio emission;

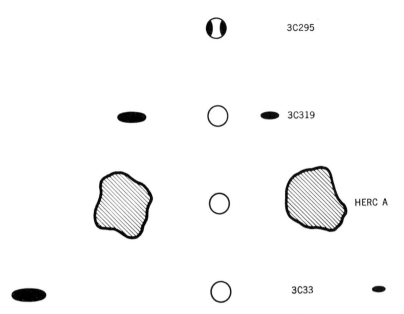

Fig. 13.1 Schematic diagrams of four radio galaxies, the radio source components being black or hatched, the galaxy a circle. These suggest an evolutionary sequence from the top to one or other of the two bottom configurations.

these indicate the presence of large-scale, ordered magnetic fields.

The first major step in understanding the nature of radio galaxies was the identification of their radio emission with the synchrotron process (Section 4.5). The necessary energy of the cosmic-ray electrons is about 10^9 eV and magnetic field strength roughly 10^{-5} gauss. The minimum total energy of these electrons and the magnetic field in which they gyrate ranges up to about 10^{60} erg. If cosmic-ray protons are also present in the proportions seen near the earth, then the total energy ranges up to about 10^{62} erg.

The second major step in interpretation was made by Shklovskii (1960, 1962), who suggested that the twin radio sources evolved as two "plasmons" of cosmic rays and magnetic field. These were ejected from the parent galaxy and expanded as they moved away in opposite directions. Their expansion causes a rapid decrease in luminosity because of decreasing cosmic-ray energy and decreasing field strength. Assuming a uniform spherical cloud of radius R, field strength B and electron density N with energy distribution

$$N(E) = KE^{-\gamma} \tag{13.1}$$

the luminosity due to synchrotron emission is

$$L \propto R^3 K B^{(\gamma+1)/2} \qquad (13.2)$$

This relationship holds for frequencies above the range where self-absorption is important. At lower frequencies, L is limited by the condition that the brightness temperature of the source cannot exceed the equivalent temperature of the cosmic-ray electrons whose motions provide the emission. In this case we have $L \propto R^2 B^{-1/2}$ and the source brightness $I \propto LR^{-2} \propto B^{-1/2}$ depends only on the strength of the magnetic field. At higher frequencies where equation 13.2 holds, the luminosity may be found from the conservation of magnetic flux, the first adiabatic invariant (for the relativistic case, with mass proportional to energy) and the conservation of the total number of cosmic-ray electrons. During expansion of a plasmon these give, respectively,

$$B \propto R^{-2} \qquad E \propto R^{-1} \qquad K \propto R^{-(2+\gamma)} \qquad (13.3)$$

which used in equation 13.2 give

$$L \propto R^{-2\gamma} \qquad I \propto R^{-2(\gamma+1)} \qquad (13.4)$$

The mean value of γ for double radio sources is 2.4 so that for the plasmon model $L \propto R^{-4.8}$. Evolutionary trends are revealed by, and may be interpreted from L–R or L–I diagrams such as those plotted by Shklovskii. A more detailed discussion of plasmon theory has been given by van der Laan (1966).

These early attempts to understand radio sources were concerned only with radio galaxies. However, the quasars have very similar radio properties (provided their redshift is of cosmological origin) and are open to a similar interpretation. Their differences in optical emission are discussed in the following section.

Another major step in evolutionary interpretation was made by Ryle and Longair (1967) who took account of the fact that in general one component of a pair of radio sources has a velocity component away from the observer and the other towards the observer. Radiation received from the receding (more distant) component must have been emitted earlier than that from the approaching component, and so the former appears brighter and closer to the parent optical object. By assuming that the angle between the line of sight and the line joining the source components has the median value of 60°, it is possible to determine the ages, velocities and displacements of the components. In this way time and velocity scales may be introduced to the L–R or L–I diagrams and further interesting inferences are possible.

Meanwhile Shklovskii's earlier evolutionary plots had been greatly

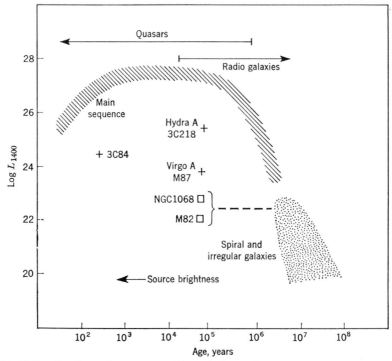

Fig. 13.2. A schematic representation of the evolutionary sequence of quasars and radio galaxies. The luminosity L (at 1400 MHz) is plotted against source brightness and age; and most galaxies and quasi-stellar objects fall within the dashed area marked main sequence. Also shown are some luminosity–brightness data for "normal" galaxies; the age scale does not apply for these.

extended by Heeschen (1966), who included quasars as well as many more radio galaxies (together these sources make up the main sequence) and some spiral and irregular galaxies. In Figure 13.2 we show an idealized summary of Heeschen's data with source luminosity (at 1400 MHz) plotted against source brightness (arbitrary logarithmic scale). A scale of source age is introduced from the results of Ryle and Longair, but it should be stressed that this applies only for quasars and radio galaxies and not for other sources whose ages cannot be determined in this way. One might also add a velocity scale to this plot, and this would show that the "main sequence" sources were ejected with velocities close to that of light and more or less brought to rest within distances of a few hundred kiloparsecs. At the same time the average source size increases from 1 kpc or less to about 100 kpc.

From these results we may draw, at least tentatively, three highly significant conclusions.

(1) Most sources fall within the hatched main sequence, which indicates that quasars and radio galaxies constitute an evolutionary sequence. The obvious difficulty of the lack of any corresponding evolution of the optical object is discussed in the following section.

(2) The initial outward motions of the sources with velocities close to c clearly differentiate this phenomenon from the violent activity seen in the central systems of many "normal" galaxies such as M82. The energy per gram of material for the relativistic ejection exceeds that for hydromagnetic ejection by factors of at least 10^3 and probably much larger. Models for the relativistic ejection are discussed in Section 13.3.

(3) The simple "plasmon" model must be abandoned, being inconsistent with the L–R relationship. During the early development of the radio sources the luminosity actually increases, but although this is a rather surprising result it is easily explained in terms of self-absorption. However, during the period 10^3–10^5 years the source size increases by a factor of about 20 and so, according to equation 13.4, the luminosity should decrease by a factor of about 10^6. In fact, the main sequence curve shows a constant value of luminosity during this period. The enormous discrepancy cannot be explained by self-absorption, but only by large increases in the magnetic energy or cosmic-ray energy (or both) of the source. A model which provides for this increase is described in Section 13.3.

The relationship between the main sequence sources and the spiral and irregular galaxies (which include the two misfits NGC1068 and M82) is discussed in the following section. The other three misfits are so-called "cores" of core-halo distributions, each being small compared to galactic dimensions and apparently centered in a galaxy. On the available data it is not clear to which sequence (if either) these radio sources belong.

13.2 Quasars and Quasi-Stellar Objects

The first radio source to be identified with a starlike object (QSO) was 3C48, and the identification was made simply because the accuracy of location of the radio source (about $\pm 5''$) eliminated all but this one stellar object. When the object was found to have a faint wisp of nebulosity, an excess of ultraviolet radiation and broad emission lines which were unlike those of any star, the identification was certain.

Other identifications were made by the same method, and other starlike objects of similar characteristics were found with no detectable radio emission (radio-quiet QSOs).

The real implications of this discovery only became evident when the spectrum lines were shown to be normal lines redshifted by very large amounts. Redshifts are known for more than 100 QSOs and values of $\delta\lambda/\lambda = z$ range from $z \sim 0.16$ to $z \sim 2.4$. Attempts have been made to explain these shifts in terms of very powerful gravitational fields at the source, and in terms of Doppler shifts in nearby objects all moving away from our galaxy. However, it is generally believed that the redshift is of cosmological origin, and on this straightforward Hubble interpretation the QSOs lie at distances up to about 3000 Mpc, or nearly the radius of the observable universe. This interpretation meets difficulties in explaining the very large and time-varying power output, but these are reduced by observations of time variations in compact galaxies, and the interpretation is adopted here.

At these distances QSOs have enormous absolute luminosities emitting 100 times or more light than normal galaxies (up to 10^{46} erg/sec). In the case of quasar 3C273 the maximum emission is in the infrared and millimeter range (10^{11}–10^{13} c/sec), the total flux being 2×10^{47} erg/sec. As seen in the preceding section, quasars and radio galaxies have rather similar radio properties. In the earlier surveys they were indistinguishable, but we now know that quasars often have at least one radio source component that is very small. The power outputs in the radio spectrum are comparable although, as we have seen, the older radio galaxies have become much weaker.

One of the remarkable features of the optical and radio emissions of quasars is their time variations. Observed changes in radio emission depend on wavelength and on the particular quasar, but large changes within a period of a year or so have been recorded. Optical changes are even more remarkable, observed time scales being as short as a few days. The significance of such changes is that one would expect the source size to be no greater than the time scale, τ multiplied by the velocity of light, c. For the optical source we would have $\tau c \sim 10^{16}$ cm, smaller by a factor of 10^6 than a small galaxy; the radio source size would be limited to about 10^{18} cm or a fraction of a parsec. Such sizes seem ridiculously small in view of the power outputs, and this has led to suggestions that the objects are not at cosmological distances but are local objects of much lower luminosities. Other ways around the size difficulty have been suggested by Rees and Simon (1968) who invoke relativistic motions within the sources, and by Morrison and

Sartori (1968) who invoke a focusing effect which greatly magnifies the rate of variation in a particular direction of emission.

The implications of the quasar measurements were so drastic that they prompted a series of interdisciplinary symposia under the heading "relativistic astrophysics." The earlier discussions have been edited by Robinson et al. (1965) and more recent ideas reviewed by Maran and Cameron (1967). Comprehensive reviews of quasar observations and their interpretations have been given by Burbidge (1967) and Burbidge and Burbidge (1967). From the point of view of cosmic electrodynamics, quasars may provide three main problems.

(1) The twin radio sources may originate in a manner similar to that of radio galaxies. The relationship was discussed above and a model is described in the following section.

(2) The optical emission comprises broad lines (up to 100 Å wide) apparently originating at temperatures about 3×10^4K. There is also continuum radiation which may extend into the infrared and milli-meter range. In these and other respects the QSOs show a close resemblance to Seyfert nuclei, except for a much greater power output. As seen below, the time scale for QSOs may be about 10^6 years in which case their total power output would be 10^{60} erg. This "magic" figure seems to apply for quasar radio sources, radio galaxies and Seyfert nuclei as well as QSOs. This is encouraging to theoreticians because it is most unlikely that there is more than one such enormous energy source. This must surely be gravitational energy from a single large mass, and must somehow be channeled into the various forms observed.

Partly on this basis we propose a model of QSOs similar to that for Seyfert nuclei (Section 12.4). A mass in the range 10^8–10^{10} $M_\odot$ is magnetically braked until it is able to release gravitational energy in the forms required. For reasonable initial dimensions the time required to remove all angular momentum (equation 12.7) may be 10^9 years or more. However, during most of this period gravitational energy is converted mainly to rotational kinetic and magnetic energies. During a critical period, which may be short, support of the mass is transferred from inertial force to gas and magnetic pressure and energy is transformed to forms which provide the optical emission. The difference between Seyfert nuclei and QSOs results from the pattern of magnetic braking. In the former, successive, thin plasma shells are braked to rest and collapse; in the latter, the whole mass reaches a critical stage and collapses within a much shorter period.

An electromagnetic theory of QSOs has been advanced by Sturrock and Feldman (1968) similar to their theory of Seyfert nuclei discussed in Section 12.4. The rate of accretion required is very much greater for QSOs and appears to pose a major difficulty.

(3) The relationship of QSOs to optical galaxies, both of the main sequence and others of Figure 13.2, provides some problems. There do not appear to be any QSOs associated with radio sources older than about 10^6 years; this suggests that QSOs may have lifetimes limited to this span, after which the radio sources join the list of unidentified sources. More difficult to answer is the question of why there seem to be no radio galaxies with radio sources younger than about 10^4 years. The answer may be statistical: relatively few should be seen during a period which is only 1% or so of the total life of the sources. It is possible that there are also real differences, as discussed in the following section.

The final question is the relationship between the main sequence and the spirals and irregulars in the luminosity–brightness plot of Figure 13.2. In spite of some indication of continuity this cannot be an evolutionary sequence; both the optical galaxies and the radio sources have characteristics which preclude such a sequence. It would seem that the "spiral and irregular sequence" become synchrotron radio sources through a different form of activity—one which does not involve the ejection of twin clouds of cosmic rays. Perhaps M82 and NGC1068 will evolve to the spiral and irregular sequence as shown by the dashed line. In any case, the violent events in the central regions of these galaxies do not seem to be related to quasars or radio galaxies. This evidence, together with the quite different velocity scales, have led us to conclude that the former is a hydromagnetic outburst, perhaps a Rayleigh-Taylor instability, while the latter is an escaping "bubble" or stream of cosmic rays.

The theory of radio galaxies and quasars outlined above has been criticized by Burbidge and Burbidge (1967) on two counts as follows.

(1) The optical galaxies concerned are old ($\sim 10^{10}$ years), but it is unreasonable to suppose that the central plasma cloud, which is supposed to give rise to the radio source, has remained in a state of activity for the life of the galaxy.

This criticism is not valid. The very essence of the theory is that gravitational energy is slowly and steadily converted first to kinetic energy and then to magnetic energy. Only when the latter develops to a level where magnetic stresses are unsupportable does violent activity occur. In this and the preceding chapter time scales for this development are determined and found of the right order.

(2) The theory is incomplete in that it does not describe how a massive

central region could form by condensation, or why the gravitational energy given up in this process is only to be released by the types of plasma instability proposed.

This criticism is valid, but applies equally to all other attempts to understand the formation of protogalaxies. In fact, the large-scale magnetic field involved may be the very component required to overcome the difficulties of fragmentation, loss of angular momentum and so on. The numerous theories (supernova, stellar collisions, massive superstars, etc.) which ignore the role of a magnetic field are even less complete since the field is needed to provide the synchrotron emission. As far as the type of plasma instability is concerned, the Rayleigh-Taylor has been investigated in some detail in connection with galaxies and appears likely to be most important even for weaker fields than those of the model.

13.3   Origin of Twin Radio Sources

As we have seen above, there appear to be three different catastrophic phenomena which occur in the central systems of galaxies. In each case the energy requirement exceeds 10^{60} erg, although the time scales appear to range from about 10^6 years or less to perhaps 10^9 years or more. In each case gravitational energy must be invoked, but further theoretical progress requires a mechanism for transforming this to magnetic and thence to cosmic-ray energy. The role of an intergalactic field has been discussed by Ginzburg (1964) and by Piddington and other references given in Chapter 12. The spiral-field model (Piddington, 1966) has been applied to explain radio galaxies with twin sources. Only this model and the oblique field model seem capable of explaining the many complexities of galactic dynamics in terms of magnetic stresses.

The spiral-field model was proposed as an explanation of explosions in the central systems of galaxies, with particular reference to radio galaxies and quasars. It now seems that activity in central systems takes several forms, and that the spiral-field model, as far as it was developed in 1966, provides an explanation of only one. This is the hydromagnetic outburst as observed in M82, resulting from Rayleigh-Taylor instability of the tightly wound spiral field (Figure 12.6).

As seen above, the model may be further developed to explain a second form of activity: magnetic braking to a stage that allows the formation of a superstar, which may correspond to the observed optical sources in Seyfert galaxies and QSOs.

It may be of interest to attempt further development to explain

twin radio sources as seen in radio galaxies and quasars. The main
clues are the ejection velocity (the velocity of light) and the large
increase in the source energy (magnetic plus cosmic-ray) after the source
has left the parent galaxy. If the ejection involved a substantial mass
of plasma, as in the case of M82, then the energy requirement of the
whole system would be increased to about 10^{62} erg or more. Such an
increase raises difficulties which may be avoided if the twin sources
comprise jets or bubbles of cosmic-ray gas ejected into a pre-existing
magnetic field configuration. Such a model seems to warrant a brief,
qualitative discussion.

In Figure 13.3 we show schematically two stages in the development
of a pair of twin synchrotron sources. The intergalactic field B is
pinched into an hour-glass configuration in the manner shown in
Figure 12.3b. Differential rotation ω about the axis perpendicular to
the field causes a spiral field to develop internally according to
equation 12.2. Eventually this field will become strong enough to
cause a Rayleigh-Taylor instability and ejection of plasma and field

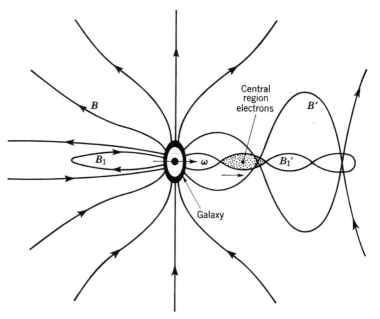

Fig. 13.3. An elliptical system with rotation vector ω perpendicular to the (frozen-
in) intergalactic magnetic field. On the left-hand side a magnetic force tube B_1
has formed by Rayleigh-Taylor instability within the galactic central region. On the
right-hand side the intergalactic field and the ejected field have been wound into
helical forms B' and B_1', and a cloud of cosmic-ray electrons ejected into the helix.

on either side of the galaxy. This ejection is depicted by the loop B_1 on the left-hand side only, so that the galaxy has developed a magnetic corona whose field strength may greatly exceed that of the original pinched field. Up to this point we have ignored the effects of galactic rotation on the external fields B and B_1. Both of these will be wound into helical forms in the region near the rotational axis, as shown schematically by the lines B' and B_1' (on the right-hand side only). Near the galactic equatorial plane the forms will tend to spirals, which are not drawn.

The winding is much faster near the rotational axis, the angular velocity at a distance of say 100 pc from the axis exceeding that at 10 kpc by a factor of more than 100. The external field near the axis is wound up more rapidly and we have a tendency towards a force-free configuration (Section 2.2) and a stronger field in this region. Now according to the virial theorem, the total magnetic energy which may be contained gravitationally must be less than the (negative) gravitational energy of a galaxy. Using equations 12.6 this limits the strength of the field B within the galaxy and the total magnetic flux $\psi = \pi R^2 B$ through the galaxy, to values

$$B^2 < 6GM^2 R^{-4} \qquad\qquad \psi < 10^3 M \qquad\qquad (13.5)$$

A galaxy of mass $10^{11}\, M_\odot$ might contain a flux of about 10^{41} gauss cm^2, and such a flux uniformly distributed over a sphere of radius 100 kpc has a field strength of only about 10^{-7} gauss. This appears to be considerably less than the fields of twin radio sources and poses a difficulty for theories based on a simple, pinched intergalactic field. However, the ejection of magnetic loops such as B_1 and the external winding up effect may resolve the difficulty.

Meanwhile, winding of the field has also proceeded within the galaxy to form tighter and tighter spirals (Figure 4.1) and more unstable magnetic neutral sheets. It is envisaged that these are triggered to provide a central region flare with the ejection of a cloud or a stream of cosmic rays along the rotational axis, as shown. The possibility arises here that the optical object in a quasar may aid in triggering the flare, and that there may be a real difference between the twin sources of quasars and radio galaxies. The former may be observable as bubbles of cosmic-ray gas, while the matter may emerge as streams from which gas clouds form. The stream hypothesis is supported by observed bridges between many pairs of sources.

There are still many difficulties to be overcome by the above, or any other, model of twin sources. First, the emerging cosmic-ray stream will tend to be arrested by the two-stream instability in its passage through the ambient plasma (Section 3.4). However, this instability tends to be suppressed as quasi-relativistic electrons join the ambient plasma.

Second, we must explain the existence of the small, discrete and widely separated sources such as in 3C33 (Figure 13.1). Somehow the clouds of electrons forming the sources must be sealed off and allowed to move and to expand as discrete objects. One way in which sealing off could be effected is by reconnection of the field line B_1' across the neutral sheet; the line then becomes a closed loop and seals in its cosmic rays. De Young and Axford (1967) explain the small size and wide separation of some twin sources in terms of ram-jet action of the interplanetary plasma. However, such action would produce shell-type sources elongated in directions perpendicular to the line joining them, rather than along the line as in Figure 13.1. Nevertheless, when there is more agreement on the nature of the sources their influence by ram-jet action may allow determinations to be made of the inter-galactic gas density.

13.4   The Intergalactic Medium

The metagalaxy, or part of the universe which can be observed with optical, radio and x-ray telescopes, has an extent of a few thousand megaparsecs and a volume of about 4×10^{84} cm^3. Within this volume there are a few times 10^{10} galaxies so that each galaxy has an average surrounding volume of a few times 10^{74} cm^3. If we consider a galaxy (including a corona) as a sphere of radius 10 kpc, then it has a volume of about 10^{68} cm^3 and occupies about one-millionth of its share of space. However, galaxies are not uniformly distributed through this space but have a strong tendency to cluster. Our Local Group of galaxies has about twenty members, but the average number is about 200 within a region of extent about 10^{25} cm so that each galaxy occupies a space only about 10^4 times its own volume. Larger clusters include those in Virgo (2500 galaxies) and perhaps the Local Supergalaxy of which the Local Group is a member. This supergalaxy has dimensions 6 Mpc $\times$ 30 Mpc, being a flattened disk.

Like the interstellar medium, the intergalactic medium comprises gas, magnetic field, cosmic rays and electromagnetic radiation. A proportion of each of these was left over when the galaxies were formed, some was subsequently ejected from galaxies, and some was developed in intergalactic space. A knowledge of these quantities is most important in theories of the origin of the universe but unfortunately, apart from the radio sources themselves, observational data are sparse.

Nevertheless, some exciting discoveries have been made recently and it seems that our understanding of intergalactic space and of the universe as a whole is increasing rapidly after a long period of stagnation. We will briefly review these and earlier results.

Intergalactic Gas

The intergalactic gas, like the interstellar gas, varies from place to place in density as well as other qualities. Direct measurement has not been possible, but the density within our Local Group required to provide stability is about 5×10^{-28} g/cm^3 (Kahn and Woltjer, 1959). Recently the interpretation of a background of soft x radiation has led Henry et al. (1968) to a density of about 10^{-29} g/cm^3 for the meta-galaxy. This is satisfactorily close to the crucial "cosmological" value discussed in the following section. The temperature inferred from the x-ray observations is a few times 10^5K, which is in agreement with the theoretical determination of Ginzburg and Ozernoi (1965) based on heating by exploding galaxies and other effects. The high temperature also explains the failure to observe other effects of the intergalactic gas, notably optical and radio wave absorption by neutral hydrogen or molecular hydrogen. The energy density of the hot gas is about 10^{-15} erg/cm^3.

While the main constituent of the intergalactic gas must be hydrogen, some helium will be present and the proportion may be estimated roughly by observations of various local objects. These show proportions of He which seem to be too large to be accounted for by nucleosynthesis in stars followed by ejection from the stars (see Tayler, 1967, and others in this reference). This difficulty would be overcome if primaeval matter contained a substantial proportion of He instead of being pure H. Furthermore, the existence of such He is readily accounted for by Gamow's universal fireball and so the excess He provides evidence in favor of that theory (Section 13.5). A crucial test is that there is a substantial amount of intergalactic He, and intimately connected to this question is the intergalactic radiation field discussed below.

Intergalactic Magnetic Field

The intergalactic magnetic field seems to be essential to explain the origin of galactic magnetic fields. Other evidence is found in the configurations of some radio galaxies which depart from the idealized forms shown in Figure 13.1. The curious convolutions of Centaurus A and a number of the sources whose brightness distributions are given by Macdonald *et al.* (1968) are highly suggestive of magnetic fields and their asymmetric stresses. Similar conclusions may be drawn from the optical brightness distributions of some galaxies, notably the bent galactic disks and the bridges between galaxies. The magnetic theory of galactic forms, spiral arms and central activity discussed in Chapter 12 depends, of course, on an intergalactic field.

A galactic field is formed from an intergalactic field by the contraction of a gas cloud of dimensions about 300 kpc to about 30 kpc to give a field of strength about 10^{-7} gauss. Subsequent winding by differential rotation is able to account for the much stronger galactic fields observed or inferred. The strength of the frozen-in field in the original gas cloud must be about 10^{-9} gauss and the cloud density a few times 10^{-28} g/cm^3 which is the density inside a galactic cluster. If we now extrapolate further to a gas density of 2×10^{-29} g/cm^3, the field strength must be reduced further to a few times 10^{-10} gauss. The energy density of this "cosmological" field is a few times 10^{-21} erg/cm^3.

When a general reluctance to accept an intergalactic magnetic field is overcome, it may also be of assistance in explaining the apparent short time scale of the self-gravitation to form galaxies, and also their large random motions. Its possible contribution in cosmology is discussed in the following section.

Intergalactic Cosmic Rays

Intergalactic cosmic rays have been proposed as the main source of primary radiation seen near the earth. However, as seen in Section 12.4 an energy density of 10^{-15} erg/cm^3 seems optimistic; one of 10^{-18} erg/cm^3 may be more realistic. This figure lies between the estimated energy densities of the hot gas and the weak magnetic field.

Although it seems probable that most of the primary cosmic radiation originates within our galaxy, this does not apply to the very energetic particles (above 10^{15} eV) which are presumably accelerated by the Fermi process in intergalactic space as a consequence of changes in the magnetic field configuration.

Intergalactic Electromagnetic Radiation

Intergalactic electromagnetic radiation probably includes the isotropic black body radio flux. Penzias and Wilson (1965) compared their 7.4 cm aerial temperature with the Johnson noise of a resistive termination immersed in liquid helium. Allowing for atmospheric and ground emission and pickup from the circuitry, a field of about 3K remained. The black body form of this radiation has been shown by measurements in the range 2.6 mm to above 50 cm.

The question of whether this radiation is local or universal is of vital importance in connection with the origin of the universe (Dicke *et al.*, 1965). In this way it is closely related to the question of helium abundance discussed above and in Section 13.5. A direct test is difficult but not impossible because of the effect of the radiation on cosmic rays. There should be a sharp cut-off of gamma rays with energy above 10^{15} eV and of particles with energy above 10^{20} eV. No conclusive results are available but Rossi's group have detected one particle above 10^{20} eV.

On the theoretical side we note that the energy density of 3K black body radiation is about 10^{-12} erg/cm³ which, as Hoyle has pointed out, is just the same as that of galactic cosmic rays and magnetic field. This is much higher than the energy densities of any of the other constituents, which may seem surprising. Nevertheless, the isotropy of the radiation and its black body spectrum seems to preclude a more local origin.

For a more comprehensive review of the intergalactic medium and the processes operating, the reader is referred to Gould (1968).

13.5 The Magnetic Universe

Until relatively recently cosmology was mainly a matter of geometry, kinematics and to some degree dynamics as formulated in Einstein's general theory of relativity. Now we are becoming more interested in "physical" cosmology, or the physical processes mainly responsible for making the universe as it is. These include nucleosynthesis, the interaction of particles and quanta with one another, and electrodynamic effects. A review of some of these advances is given by Novikov and Zeldovic (1967). It is also interesting to note that some discoveries which stimulated these new interests (quasars, the 3K black body radiation, the He excess and so on) have made new demands on the general theory of relativity. As shown below, the effects predicted by this theory in stars and stellar systems are extremely small and it has been possible to detect and measure only three, including the curvature of light rays in the sun's gravitational field. However, if the energy of quasars and radio galaxies derives from gravitational energy in the superstar or galactic central region, then the theory will find application there.

In 1948 Gamow showed that if the universe started as a hot "Big Bang," then at an epoch t sec the black body temperature should, on certain assumptions, be given by

$$T = 2 \times 10^{10} t^{-1/2} \text{K} \qquad (13.6)$$

This temperature is inferred from the spectrum and energy density of the electromagnetic radiation which has become dissociated from the matter of the universe. For $t = 3 \times 10^{17}$ sec (the Hubble red-shift age of 10^{10} years), $T = 36K$. For many years no attempt was made to check this prediction but, as seen above, black body radiation has been detected at a somewhat lower level. According to recent advances in the theory of nucleosynthesis, the universal fireball might also explain the observed excess of helium (Dicke, 1968). These discoveries all provide evidence in favor of Gamow's theory, but there remains a major difficulty in the form of a "singularity" discussed below.

At the present time (as measured by clocks on the earth) the universe is expanding and there would appear to be two possible future developments. If its average mass density is low enough, then the universe is "open," and matter and radiation may escape and move apart indefinitely. Such a cosmology carries with it the implication that matter originated at some finite time in the past and here we meet the difficulty that one would expect matter and antimatter to have been created in equal quantities whereas the former greatly predominates. For this and other reasons this version of the Big Bang is aesthetically displeasing and its alternative is favored.

The alternative is that the average density of the universe exceeds 2×10^{-29} g/cm³ (the "cosmological value"), in which case the universe will eventually stop expanding and suffer gravitational collapse. In its present form this version offers little advantage, because the cosmological solutions of Einstein's field equations lead to a "singularity." Spherical gravitational collapse terminates in a region of space–time where infinitely strong gravitational forces deform matter and photons out of recognition and squeeze them out of existence, perhaps to reappear elsewhere in the space-time continuum. This requirement for the re-creation of all matter meets the difficulty mentioned above, and so strenuous efforts are being made to explain away such a fate for the universe. In these, electrodynamic effects may be important.

Before briefly discussing these efforts we must comment on the apparent paradox that gravity is so important in cosmology and yet such a negligible force in the laboratory, and even smaller on an atomic or nuclear level. For example, the electromagnetic forces holding atoms and molecules together are larger by factors of about 10^{37}. Even at the level of stars and stellar systems the general theory of relativity predicts only small effects; a photon passing near the sun is deflected only about 1.7 sec of arc. At the level of galactic central systems and the universe, however, where huge numbers of atoms combine their pull, gravity becomes the one irresistible force. An important aspect of this force, and the available energy, is revealed by equation 12.6. If we reduce the radius (or increase the mass) of a spherical system until the gravitational energy released is equal to half the rest-mass energy of the system ($\Omega = 1/2Mc^2$), then we have reached the radius of gravitational collapse

$$R_g = 2GMc^{-2} \tag{13.7}$$

at which gravity becomes an irresistible force and destroys the system. In the case of the earth, for example $R_g \sim 0.9$ cm. In the case of the universe, R_g equals the present radius for a density equal to the cosmological value.

A number of attempts have been made to avoid the "singularity" in the solution of Einstein's field equations. As in the early models of the atom, where a similar difficulty was met, we may be forced to abandon the continuous, deterministic nature of space–time and adopt a quantum theory of gravity. However, it is possible that less drastic assumptions may serve. One of these is to introduce a general rotation, but so far this has not been successful. Another possibility is Dicke's zero-mass scalar-field (introduced to limit helium production) or perhaps a magnetic field, which adds to the energy but not to the mass.

Let us suppose then that magnetic or other forces have eliminated the singularity so that the universe oscillates or bounces in and out instead of collapsing. It is a closed universe from which no matter or photons may escape but, as we shall see, the magnetic energy content may change with drastic effects. The matter that we see now represents the same baryon content of the previous expansions and so avoids the question of the origin of matter at any finite time in the past. Each contraction must proceed to a temperature of 10^{10}K or more (radius about 1 pc) when all of the ashes of the previous cycle are reprocessed back to the hydrogen and helium needed for the next cycle. The black body radiation of the fireball cannot escape, but is adiabatically cooled by the cosmological redshift while preserving its thermal character, the temperature varying inversely as the radius.

Even if such a cosmology is accepted, there remain many problems including the origin of galaxies. This would seem to require substantial density perturbations when the universe was very young and, unless these were carried over from the preceding contraction, they may require the presence of an irregular force field. Other problems are posed by the QSOs which are seen at distances near the limits of the visible universe (for a review of radio source counts, see Ryle, 1968). The optical line emission indicates element abundances similar to those in our Population 1 objects (young stars) which are made out of old "well cooked" gas. These two effects seem to indicate that the objects which were later to become radio and optical QSOs developed early in the history of the universe. They developed stars in which the primaeval gas was "cooked" and suffered explosions a few times 10^9 years ago, while our Local System was relatively young.

We will conclude by considering briefly the possible origins and effects of a cosmological magnetic field. If there is differential rotation of the universe, then a "seed" field, however weak, will develop (equation 12.2) during oscillation after oscillation and so may account for the present field of strength perhaps a few times 10^{-10} gauss. In an oscillating universe another contribution must be made from the magnetic fields of radio galaxies, whatever the origin of the original galactic fields themselves. We will make a rough estimate of their contributions on an energy basis (rather than a magnetic-flux basis, because there are a large number of isolated systems). Assuming 10^2 generations of radio sources in the present expansion, with 10^5 sources per generation and 10^{56} erg per source, we have total magnetic energy 10^{63} erg and an average energy density in the universe of 10^{-21} erg/cm^3. The present estimate of magnetic energy density is only a few times larger, which suggests that during each oscillation of the universe the magnetic energy is increasing. A requirement, of course, is that the electrical conductivity is sufficiently high, in which case the universe is evolutionary as well as oscillatory.

Some possible effects of an evolving field are the following:

(1) It may, perhaps in combination with differential rotation, prevent complete collapse to a singularity. The effect of the magnetic field, as estimated from a comparison of its total energy with the (negative) gravitational energy, is admittedly small, but its control may be more subtle as in the case of galactic dynamics.

(2) The field may play an important part in the genesis of galaxies or of galactic clusters. The requirement here is an initial perturbation and the law of chance is usually invoked to start a condensation (Doroshkevich et al., 1967). A magnetic field introduces anisotropy, neutral sheets and Rayleigh-Taylor instability and so may initiate cloud formation. Since any magnetic field will continue to evolve from cycle to cycle, previous cycles may have been free of galaxies and later cycles may produce new forms.

(3) Magnetic neutral sheets may be a source of intergalactic cosmic rays.

References

Burbidge, E. M., 1967, *Ann. Rev. Astron. Astrophys.* **5**, 399.

Burbidge, G. R. and Burbidge, E. M., 1967, *Quasi-Stellar Objects*, Freeman, New York.

De Young, D. S. and Axford, W. I., 1967, *Nature (London)* **216**, 129.

Dicke, R. H., 1968, *Astrophys. J.* **152**, 1.

Dicke, R. H., Peebles, P. J. E., Roll, P. G. and Wilkinson, D. T., 1965, *Astrophys. J.* **142**, 414.

Doroshkevich, A. G., Zeldovich, Ya. B. and Novikov, I. D., 1967, *Soviet Astron. AJ* **11**, 233; *Astron. Zh.* **44**, 295.

Ginzburg, V. L., 1964, *Soviet Phys. Doklady* **9**, 329.

Ginzburg, V. L. and Ozernoi, L. M., 1965, *Soviet Astron.* **9**, 726; *Astron. Zh.* **42**, 943.

Gould, R. J., 1968, *Ann. Rev. Astron. Astrophys.* **6**, 195.

Heeschen, D. S., 1966, *Astrophys. J.* **146**, 517.

Henry, R. C., Fritz, G., Meekins, J. F., Friedman, H. and Byram, E. T., 1968, *Astrophys. J.* **153**, L11.

Hoyle, F. and Fowler, W. A., 1963, *Mon. Not. Roy. Astron. Soc.* **125**, 169.

Kahn, F. D., and Woltjer, L., 1959, *Astrophys. J.* **130**, 705.

Macdonald, G. H., Kenderline, S. and Neville, A. C., 1968, *Mon. Not. Roy. Astron. Soc.* **138**, 259.

Maran, S. P. and Cameron, A. G. W., 1967, *Science*, **157**, 1517.

Moffet, A. T., 1966, *Ann. Rev. Astron. Astrophys.* **4**, 145.

Morrison, P. and Sartori, L., 1968, *Astrophys. J.* **152**, L139

Novikov, I. D. and Zeldovic, Ya. B., 1967, *Ann. Rev. Astron. Astrophys.* **5**, 627.

Penzias, A. A. and Wilson, R. W., 1965, *Astrophys. J.* **142**, 419.

Piddington, J. H., 1966, *Mon. Not. Roy. Astron. Soc.* **133**, 163.

Rees, M. J. and Simon, M., 1968, *Astrophys. J.* **152**, L145.

Robinson, I., Schild, A. and Schucking, E. L., eds., 1965, *Quasi-Stellar Sources and Gravitational Collapse*, University of Chicago Press, Chicago.

Ryle, M., 1968, *Ann. Rev. Astron. Astrophys.* **6**, 249.

Ryle, M. and Longair, M. S., 1967, *Mon. Not. Roy. Astron. Soc.* **136**, 123.

Shklovskii, I. S., 1960, *Soviet Astron. AJ* **4**, 234; *Astron. Zh.* **37**, 256.

Shklovskii, I. S., 1962, *Soviet Astron. AJ* **6**, 465; *Astron. Zh.* **39**, 591.

Sturrock, P. A. and Feldman, P. A., 1968, *Astrophys. J.* **152**, L39.

Tayler, R. J., 1967, *Observatory* **87**, 193.

van der Laan, H., 1966, *Nature (London)* **211**, 1131.

Author Index

Subject Index

301